The Pocket Atlas
of Human Anatomy

The Pocket Atlas of Human Anatomy

A Reference for Students of Physical Therapy, Medicine, Sports, and Bodywork

Chris Jarmey

lotus
p u b l i s h i n g
Chichester, England

North Atlantic Books
Berkeley, California

First published in 2018 by
Lotus Publishing
Apple Tree Cottage, Inlands Road, Nutbourne, Chichester, PO18 8RJ and
North Atlantic Books
Berkeley, California

All Drawings Amanda Williams
Text Design Medlar Publishing Solutions Pvt Ltd., India
Printed and Bound in India by Replika Press

The Pocket Atlas of Human Anatomy: A Reference for Students of Physical Therapy, Medicine, Sports, and Bodywork is sponsored and published by the Society for the Study of Native Arts and Sciences (dba North Atlantic Books), an educational nonprofit based in Berkeley, California, that collaborates with partners to develop cross-cultural perspectives, nurture holistic views of art, science, the humanities, and healing, and seed personal and global transformation by publishing work on the relationship of body, spirit, and nature.

North Atlantic Books' publications are available through most bookstores. For further information, visit our website at www.northatlanticbooks.com or call 800-733-3000.

British Library Cataloguing-in-Publication Data
A CIP record for this book is available from the British Library
ISBN 978 1 905367 85 6 (Lotus Publishing)
ISBN 978 1 62317 252 7 (North Atlantic Books)

Library of Congress Cataloguing-in-Publication Data
Names: Jarmey, Chris, author. | Society for the Study of Native
        Arts and Sciences, sponsoring body.
Title: The pocket atlas of human anatomy / Chris Jarmey.
Description: Berkeley, California : North Atlantic Books, 2018.
Identifiers: LCCN 2017052631 (print) | LCCN 2017053235 (ebook) |
        ISBN 9781623172534 (ebook) | ISBN 9781623172527 (pbk.)
Subjects: | MESH: Anatomy | Atlases
Classification: LCC QM23 (ebook) | LCC QM23 (print) | NLM QS
        17 | DDC 611—dc23
LC record available at https://lccn.loc.gov/2017052631

Contents

Foreword

Between 1988, when Chris married my sister and 2008, the year of his untimely death, I had the good fortune to benefit from Chris' depth of knowledge on matters pertaining to the health of the human body. As a physiotherapist and shiatsu practitioner, his understanding of anatomy and physiology was extensive. In addition, Chris' skills as a teacher of shiatsu, yoga and meditation were widely respected, in particular his ability to explain complex concepts in simple, practical terms.

As a practicing GP I was delighted when Chris began to write books that encompassed the wide scope of his expertise. *The Concise Book of the Moving Body*, along with its companion *The Concise Book of Muscles*, sits on the shelf in my GP surgery. It is especially helpful as an aide-memoire when I need a quick revision of anatomy. The clear, concise language and accompanying illustrations are easy to access and take the reader straight to the point. On occasion, I have found it useful in explaining symptoms to my patients, who are grateful for explanations that help demystify their ill health.

I'm delighted that *The Concise Book of the Moving Body* has now been repackaged and rebadged as *The Pocket Atlas of Human Anatomy*, and I am confident that it will serve as an excellent text for all students of anatomy and bodywork, whether in the field of medicine, sport or rehabilitation. I have no hesitation in recommending Chris' textbook and will continue to appreciate the valuable contribution he has made. My hope is that this book will enhance your understanding of the human body and will become a trusted companion on your journey of discovery.

Dr. David Simpson, 2018

A Note About Peripheral Nerve Supply

The *peripheral nervous system (PNS)* comprises all the neural structures outside the brain and spinal cord, which constitute the *central nervous system (CNS)*. The PNS has two main components: the *somatic nervous system* and the *autonomic nervous system*; the latter deals with involuntary control of smooth muscle and glands. As this book is concerned with skeletal muscles, it is only the somatic nervous system that is of interest.

The PNS consists of 12 pairs of cranial nerves and 31 pairs of spinal nerves, along with their subsequent branches. The spinal nerves are numbered according to the level of the spinal cord from which they arise, known as the *spinal segment*. Muscle innervation pathways are described in detail in Appendix 1.

In Appendix 2 the relevant peripheral nerve supply is listed with each muscle, as this information may be useful for healthcare practitioners. However, the spinal segment* from

---

*A spinal segment is the part of the spinal cord that gives rise to each pair of spinal nerves, one for each side of the body. Each spinal nerve contains sensory and motor fibers from the dorsal and ventral roots respectively. Soon after the spinal nerve exits through the foramen or opening between adjacent vertebrae, it divides into a dorsal primary ramus, which is directed posteriorly, and a ventral primary ramus, which is directed anteriorly and laterally. Fibers from the dorsal

which the nerve fibers arise often varies between different sources. This is because spinal nerves are organized into networks known as *plexuses* (plexus = a network of nerves: from Latin *plectere* = "to braid"), which supply different regions of the body, and nerve fibers from different spinal segments will contribute to the individual named nerve that supplies a particular muscle.

For each muscle in Appendix 2, the spinal levels that typically contribute to its named nerve are indicated. The relevant spinal segments are represented by C for cervical, T for thoracic, L for lumbar, and S for sacral, followed by a number representing the level.

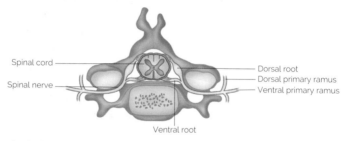

Spinal cord

Spinal nerve

Dorsal root
Dorsal primary ramus
Ventral primary ramus

Ventral root

A spinal segment, showing the nerve roots combining to form a spinal nerve, which then divides into ventral and dorsal rami.

rami innervate the skin and extensor muscles of the neck and trunk. The ventral rami supply the limbs, as well as the sides and front of the trunk.

CHAPTER 1

Anatomical Terms

Positions

To describe the relative positions of body parts and their movements, it is essential to have a universally accepted initial reference position. This is known as the *anatomical position*, which is simply the upright standing position, with feet flat on the floor, arms hanging by the sides and the palms facing forward (see Figure 1.1). The directional terminology used always refers to the body as if it were in the anatomical position, regardless of its actual position. Note also that the terms *left* and *right* refer to the sides of the object or person being viewed, and not those of the reader.

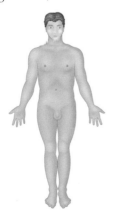

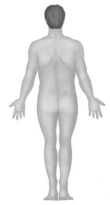

Figure 1.1: Anterior. In front of; toward or at the front of the body.

Figure 1.2: Posterior. Behind; toward or at the back of the body.

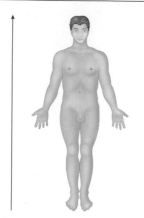

Figure 1.3: Superior. Above; toward the head or the upper part of the structure or the body.

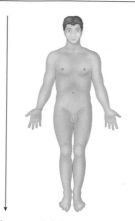

Figure 1.4: Inferior. Below; away from the head or toward the lower part of the structure or the body.

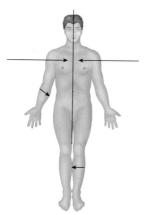

Figure 1.5: Medial. (from Latin *medius* = "middle"). Toward the midline of the body; on the inner side of a limb.

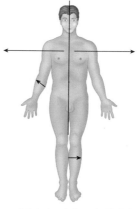

Figure 1.6: Lateral. (from Latin *latus* = "side"). Away from the midline of the body; on the outer side of the body or a limb.

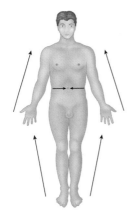

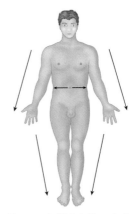

Figure 1.7: Proximal. (from Latin *proximus* = "nearest"). Closer to the center of the body (the navel), or to the point of attachment of a limb to the trunk.

Figure 1.8: Distal. (from Latin *distans* = "distant"). Farther away from the center of the body, or from the point of attachment of a limb to the trunk.

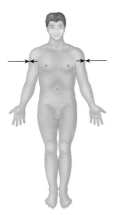

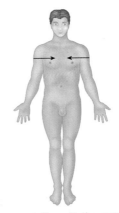

Figure 1.9: Superficial. Toward or at the body surface.

Figure 1.10: Deep. Farther away from the body surface; more internal.

Figure 1.11: Dorsal. (from Latin
dorsum = "back"). On the posterior
surface, e.g. the back of the hand.

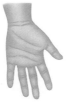

Figure 1.12: Palmar. (from Latin
palma = "palm"). On the anterior
surface of the hand, i.e. the palm.

Figure 1.13: Plantar. (from Latin
planta = "sole"). On the sole
of the foot.

Regions

The two primary divisions of the body are its *axial* parts,
consisting of the head, neck, and trunk, and its *appendicular* parts,
consisting of the limbs, which are attached to the axis of the body.
Figures 1.14 and 1.15 shows the terms used to indicate specific
body areas. Terms in parentheses are the lay terms for the area.

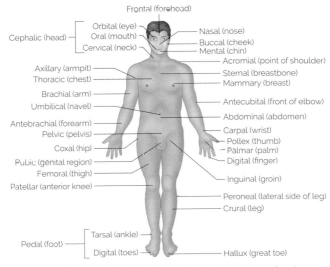

Figure 1.14: Terms used to indicate specific body areas, anterior view.

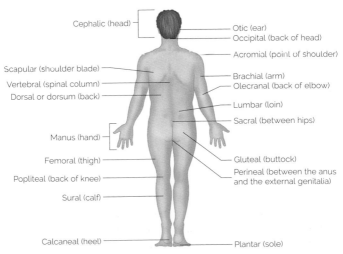

Figure 1.15: Terms used to indicate specific body areas, posterior view.

Planes

The term *plane* refers to a two-dimensional section through the body; it provides a view of the body or body part, as though it has been cut through by an imaginary line.

- The sagittal planes cut vertically through the body from anterior to posterior, dividing it into right and left halves. Figure 1.16 shows the mid-sagittal plane. A *para-sagittal plane* divides the body into unequal right and left parts.
- The frontal (coronal) planes pass vertically through the body, dividing it into anterior and posterior sections, and lie at right angles to the sagittal plane.
- The transverse planes are horizontal cross sections, dividing the body into upper (superior) and lower (inferior) sections, and lie at right angles to the other two planes.

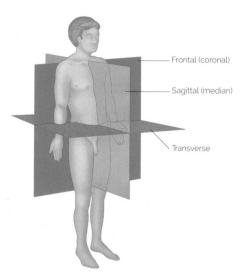

Frontal (coronal)

Sagittal (median)

Transverse

Figure 1.16: The most frequently used planes of the body.

Movements

The direction in which body parts move is described in relation to the fetal position. Moving into the fetal position results from flexion of all the limbs; straightening out of the fetal position results from extension of all the limbs.

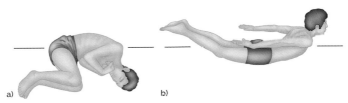

a) b)

Figure 1.17: (a) Flexion into the fetal position. (b) Extension out of the fetal position.

Main Movements

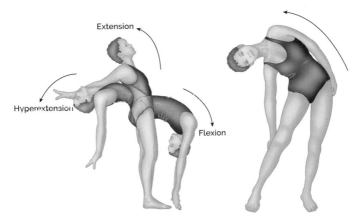

Figure 1.18: Flexion: bending to decrease the angle between bones at a joint. From the anatomical position, flexion is usually forward, except at the knee joint where it is backward. The way to remember this is that flexion is always toward the fetal position. **Extension:** to straighten or bend backward away from the fetal position. **Hyperextension:** to extend the limb beyond its normal range.

Figure 1.19: Lateral flexion: to bend the trunk or head laterally (sideways) in the frontal (coronal) plane.

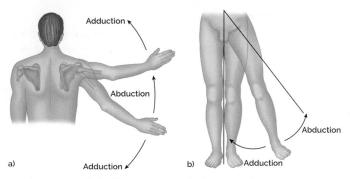

Figure 1.20: Abduction: movement of a bone away from the midline of the body or a limb. **Adduction:** movement of a bone toward the midline of the body or a limb.

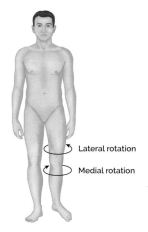

Figure 1.21: Rotation: movement of a bone or the trunk around its own longitudinal axis. **Medial or internal rotation:** to turn inward, toward the midline. **Lateral or external rotation:** to turn outward, away from the midline.

Other Movements

Movements described in this section are those that occur only at specific joints or parts of the body, usually involving more than one joint.

Figure 1.22: Pronation: to turn the palm of the hand down to face the floor (if standing with elbow bent 90 degrees, or if lying flat on the floor) or away from the anatomical and fetal positions.

Figure 1.23: Supination: to turn the palm of the hand up to face the ceiling (if standing with elbow bent 90 degrees, or if lying flat on the floor) or toward the anatomical and fetal positions.

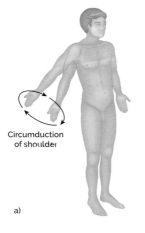

Circumduction of shoulder

a)

b)

Circumduction of leg

Figure 1.24: Circumduction: movement in which the distal end of a bone moves in a circle, while the proximal end remains stable; the movement combines flexion, abduction, extension, and adduction.

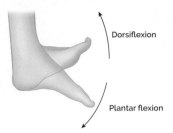

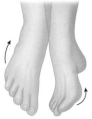

Eversion   Inversion

Figure 1.25: Plantar flexion: to point
the toes down toward the ground.
Dorsiflexion: to point the toes up
toward the ceiling.

Figure 1.26: Inversion: to turn the
sole of the foot inward, so that the
soles would face toward each other.
Eversion: to turn the sole of the foot
outward, so that the soles would
face away from each other.

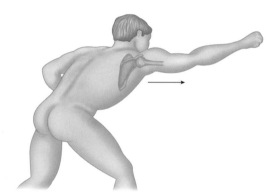

Figure 1.27: Protraction: movement forward in the transverse plane—for
example, protraction of the shoulder girdle, as in rounding the shoulder.

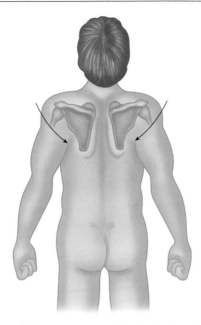

Figure 1.28: Retraction: movement backward in the transverse plane, as in bracing the shoulder girdle back, military style.

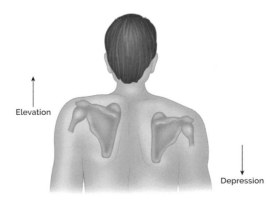

Figure 1.29: Elevation: movement of a part of the body upward along the frontal plane—for example, elevating the scapula by shrugging the shoulders. **Depression:** movement of an elevated part of the body downward to its original position.

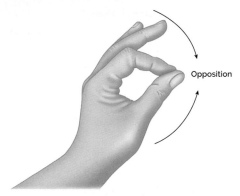

Opposition

Figure 1.30: Opposition: a movement specific to the saddle-shaped joint of the thumb; it enables you to touch your thumb to the tips of the fingers of the same hand.

CHAPTER 2

Tissues

Groups of cells having a similar structure and function are called *tissues*. There are four primary types of tissue, each having a characteristic function:

1. *Epithelial*: for example, the skin, whose function is to cover and protect the body.
2. *Connective*: for example, ligaments, tendons, fascia, cartilage, and bone, whose functions are to protect, support, and bind together different parts of the body. It is the most abundant and widely distributed tissue type.
3. *Muscle*: the function of muscle is to enable movement of the body.
4. *Nervous*: nerves control body functions and movement.

The types of tissue that concern us regarding musculoskeletal anatomy are connective tissue and muscle tissue, for which brief overviews are given below.

Connective Tissue

Common characteristics of *connective tissue* are:

1. *Variations in blood supply*. Most connective tissue is well vascularized (having blood vessels). However, tendons

and ligaments have a poor blood supply, and cartilage is avascular (having no blood vessels): therefore these structures heal very slowly.

2. *Extracellular matrix*. Connective tissues are made up of many different types of cell and varying amounts of a nonliving substance that surrounds the cells. This substance is called the *extracellular matrix*, which is produced by connective tissue cells and then secreted to their exterior. Depending on the type of tissue, the matrix may be liquid, gel-like, semisolid, or very hard.

By virtue of the matrix, connective tissue is able to bear weight and withstand stretching and other abuse, such as abrasion. The matrix contains various types and numbers of fibers—for example, collagen, elastic, or reticular.

Types of Connective Tissue

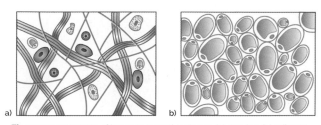

a)   b)

Figure 2.1: Structure of loose connective tissue: (a) areolar; (b) adipose.

Loose Connective Tissue

Loose connective tissue has more cells and fewer fibers; thus, it is softer than the other types. Examples are:

1. *Areolar*: a "packing" tissue, which cushions and protects body organs and holds internal organs together in their proper position.
2. *Adipose*: fat tissue that forms the subcutaneous layer beneath the skin, also called the *hypodermis* or *superficial*

fascia, where it insulates the body and protects against heat and cold.

Dense Regular Connective Tissue
Within *dense regular connective tissue*, collagen fibers are the predominant element and create a white, flexible tissue with a high resistance to pulling forces. Examples: ligaments and tendons.

Dense Irregular Connective Tissue
Dense irregular connective tissue has the same structural elements as regular connective tissue; however, the bundles of collagen fibers are thicker, interwoven, and arranged irregularly. Example: fascia.

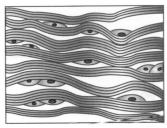

Figure 2.2: Structure of dense regular connective tissue.

Figure 2.3: Structure of dense irregular connective tissue.

Cartilage
Cartilage is tough, but flexible. It has qualities intermediate to those of dense connective tissue and bone. Cartilage is

avascular and devoid of nerve fibers, and therefore heals slowly. Examples: hyaline, fibrocartilage, and elastic.

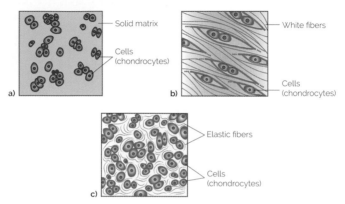

Figure 2.4: Structure of cartilage: (a) hyaline cartilage; (b) white fibrocartilage; (c) yellow elastic cartilage.

Bone
Bone cells sit in cavities called *lacunae* (sing. *lacuna*) surrounded by circular layers of a very hard matrix that contains calcium salts and larger amounts of collagen fibers.

Blood
Blood, or *vascular tissue*, is considered a connective tissue because it consists of blood cells, surrounded by a nonliving fluid matrix called *blood plasma*. The "fibers" of blood are soluble protein molecules that become visible only during blood clotting. Blood is not a typical connective tissue; it is the transport vehicle for the cardiovascular system, carrying nutrients, waste materials, respiratory gases, etc. throughout the body.

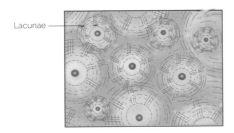

Figure 2.5: Structure of bone.

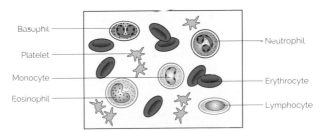

Figure 2.6: Structure of blood.

Muscle Tissue

Muscle tissue is composed of 75% water and 20% protein, with mineral salts, glycogen, and fat making up the remaining 5%. As this book is designed to focus specifically on musculoskeletal anatomy, only a brief description and comparison of the different types of muscle tissue is given below. Skeletal muscle will be discussed later in more detail (see Chapter 7).

Muscle Types and Functions
There are three types of muscle tissue: smooth, cardiac, and skeletal. All muscle cells have an elongated shape and are therefore referred to as *muscle fibers*.

Smooth/Unstriated/Involuntary Muscle

Smooth muscle cells are usually spindle shaped and arranged in sheets or layers. Smooth muscles are found in the viscera, such as the stomach, small and large intestines, blood vessels, and uterus (i.e. the hollow organs).

Smooth muscles in the blood vessels contract to move the blood in the arteries, as well as squeezing substances through the organs and tracts. These muscles are under involuntary control (although some individuals can train their minds to achieve some control over smooth muscle contractions). Contractions are usually gentle and rhythmic, with the obvious exceptions of vomiting and birth contractions.

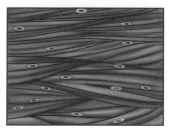

Figure 2.7: Structure of smooth/unstriated/involuntary muscle.

Cardiac/Striated/Involuntary Muscle

Cardiac muscles are found in the heart only; they exist to pump the heart and are under involuntary control. Structurally, they are made up of branching fibers that are striated in appearance and are separated or interspersed by discs known as *intercalated discs*.

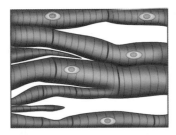

Figure 2.8: Structure of cardiac/striated/involuntary muscle.

Skeletal/Striated/Voluntary Muscle

Skeletal muscles (also called *somatic muscles*) attach to, and cover, the bony skeleton; they are under voluntary control. Skeletal muscles fatigue easily, but can be strengthened. They are capable of powerful, rapid contractions, and longer, sustained contractions. These muscles enable us to perform both feats of strength and controlled, fine movements.

Note: As they contract, all muscle types generate heat, and this heat is vitally important in maintaining a normal body temperature. It is estimated that 85% of all body heat is generated by muscle contractions.

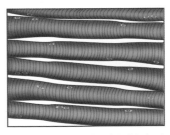

Figure 2.9: Structure of skeletal/striated/voluntary muscle.

CHAPTER 3

Bones

We are born with approximately 350 bones, but gradually some of these fuse together until puberty, when we have only 206 bones. These bones form the supporting structure of the body, and are collectively known as the *endoskeleton*. (The *exoskeleton* is well developed in many invertebrates, but exists in humans only as teeth, nails, and hair.) Fully developed bone is the hardest tissue in the body and is composed of 20% water, 30% to 40% organic matter, and 40% to 50% inorganic matter.

Bone Development and Growth

The majority of bone is formed from a foundation of cartilage (see below), which becomes calcified and then ossified to form true bone. This process occurs through the following four stages:

1. Bone-building cells, called *osteoblasts*, become active during the second or third month of embryonic life.
2. Initially, the osteoblasts manufacture a matrix of material between the cells, which is rich in a fibrous protein called *collagen*. This collagen strengthens the tissue. Enzymes then enable calcium compounds to be deposited within the matrix.
3. The intercellular material hardens around the cells, to become *osteocytes*—living cells that maintain the bone, but do not produce new bone.

4. Other cells, called *osteoclasts*, break down, remodel, and
    repair bone—a process that continues throughout life,
    but which slows down with advancing age. Consequently,
    the bones of elderly people are weaker and more fragile.

In brief, osteoblasts and osteoclasts are the cells that
respectively lay down and break down bone, enabling
bones to very slowly adapt in shape and strength
according to need.

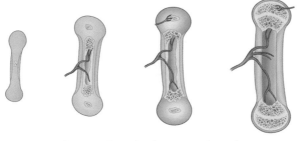

Figure 3.1: Bone development and growth.

Cartilage

Cartilage (gristle) exists either as a temporary formation that
is later replaced by bone, or as a permanent supplementation
to bone. However, it is not as hard or as strong as bone.

It consists of living cells called *chondrocytes*, contained
within lacunae (spaces) and surrounded by a collagen-rich
intercellular substance. Cartilage is relatively non-vascular
(not penetrated by blood vessels) and is mainly nourished
by surrounding tissue fluids. There are three main types of
cartilage: hyaline cartilage, white fibrocartilage, and yellow
fibrocartilage.

Hyaline Cartilage

Hyaline cartilage forms the temporary foundation of cartilage from which many bones develop, thereafter existing in relation to bone as the:

- Articular cartilage of synovial joints.
- Cartilage plates between separately ossifying areas of bone during growth.
- Xiphoid process of the sternum (which ossifies late or not at all) and the costal cartilages.

Hyaline cartilage also exists in the nasal septum, most cartilages of the larynx, and the supporting rings of the trachea and bronchi.

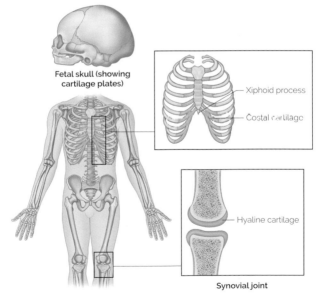

Figure 3.2: Sites of hyaline cartilage in the body.

White Fibrocartilage

White fibrocartilage contains white fibrous tissue, and has more elasticity and tensile strength than hyaline cartilage. It is found in the:

- Sesamoid cartilages in a few tendons.
- Articular discs in the wrist joint and clavicular joints.
- Rim (labrum) deepening the sockets of the shoulder and hip joints.
- Two semilunar cartilages within each knee joint.
- Intervertebral discs between adjacent surfaces of the vertebral bodies.
- Cartilage plate joining the hipbones at the pubic symphysis.

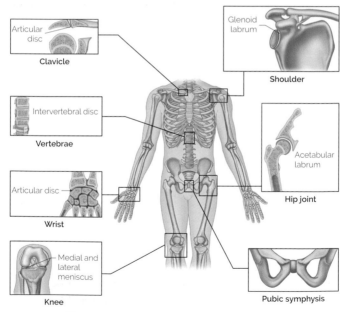

Figure 3.3: Sites of white fibrocartilage in the body.

Yellow Fibrocartilage

Yellow fibrocartilage contains yellow elastic fibers and is found in the:

- External ear.
- Auditory tube of the middle ear.
- Epiglottis.

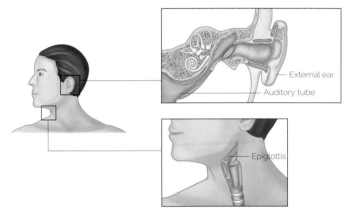

Figure 3.4: Sites of yellow fibrocartilage in the body.

Functions of Bones

Bones have several functions, including: support, protection, movement, storage, and blood cell formation.

Support

Our bones provide the hard framework that supports and anchors all the soft organs of the body. Our legs support the body's torso, head, and arms, while the ribcage supports the chest wall.

Protection
The bones of the skull protect the brain, while the vertebrae surround the spinal cord. The ribcage protects all the vital organs.

Movement
Attaching to the bones by tendons, the muscles use the bones as levers to move the body and all its parts. The arrangement of the bones and joints determines which movements are possible.

Storage
Fat is stored as "yellow marrow" in the central cavities of long bones. Within the structure of the bone itself minerals are stored. The most important minerals are calcium and phosphorus, but potassium, sodium, sulfur, magnesium, and copper are also stored. Stored minerals can be released into the bloodstream for distribution to all parts of the body as needed.

Blood Cell Formation
The bulk of blood cell formation occurs within the "red marrow" cavities of certain bones.

Types of Bone—According to Density

Compact Bone
Compact bone is dense, and looks smooth to the naked eye. Through the microscope, however, compact bone appears as an aggregation of Haversian systems, also called *osteons*. Each such system is an elongated cylinder oriented along the long axis of the bone, and consists of a central Haversian canal containing blood vessels, lymph vessels, and nerves, surrounded by concentric plates of bone called *lamellae*. In other words, each Haversian system is a group of hollow tubes of bone matrix (lamellae), placed one inside the next. Between these lamellae there are spaces (*lacunae*), which contain lymph and osteocytes. The lacunae are linked

via hair-like canals called *canaliculi* to the lymph vessels in the Haversian canal, enabling the osteocytes to obtain nourishment from the lymph. This tubular array of lamellae gives great strength to the bone.

Other canals, called *perforating canals* or *Volkmann's canals*, run at right angles to the long axis of the bone, connecting the blood vessels and nerve supply within the bone to the periosteum.

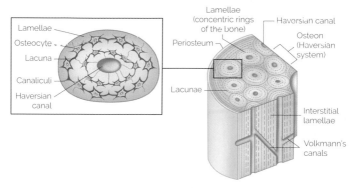

Figure 3.5: Structure of compact bone.

Spongy (Cancellous) Bone

Spongy bone is composed of small needle-like trabeculae (literally, "little beams"; sing. trabecula) containing irregularly arranged lamellae and osteocytes, interconnected by canaliculi. There are no Haversian systems but, rather, lots of open spaces that can be thought of as large Haversian canals, giving a honeycombed appearance. These spaces are filled with red or yellow marrow and blood vessels.

This structure forms a dynamic lattice capable of gradual alteration through realignment, in response to stresses of weight, postural change, and muscle tension. Spongy bone is found in the epiphyses of long bones, the bodies of the vertebrae, and other bones without cavities.

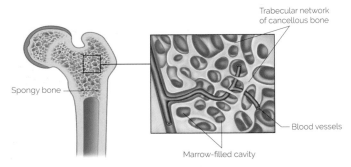

Figure 3.6: Structure of spongy (cancellous) bone.

Types of Bone—According to Shape

Irregular Bones
Irregular bones have complicated shapes; they consist mainly of spongy bone enclosed by thin layers of compact bone. Examples: some skull bones, the vertebrae, and the hipbones.

Flat Bones
Flat bones are thin, flattened bones and are frequently curved; they have a layer of spongy bone sandwiched between two thin layers of compact bone. Examples: most of the skull bones, the ribs, and the sternum.

Short Bones
Short bones are generally cube shaped and consist mostly of spongy (cancellous) bone. Examples: the carpal bones in the hand, and the tarsal bones in the ankle.

A *sesamoid bone* (from Latin *sesamoides* = "shaped like a sesame seed") is a special type of short bone that is formed and embedded within a tendon. Examples: the patella (kneecap) and the pisiform bone at the medial end of the wrist crease.

Long Bones

Long bones are longer than they are wide; they have a shaft with heads at both ends, and consist mostly of compact bone. Examples: the bones of the limbs, except those of the wrist, hand, ankle, and foot (although the bones of the fingers and toes are effectively miniature long bones).

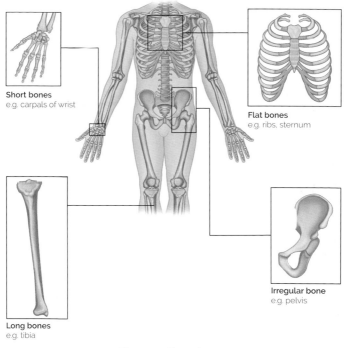

Short bones
e.g. carpals of wrist

Flat bones
e.g. ribs, sternum

Irregular bone
e.g. pelvis

Long bones
e.g. tibia

Figure 3.7: Bone shapes.

Components of a Long Bone

The transformation of cartilage within a long bone begins
at the center of the shaft. Secondary bone-forming centers
develop later on, across the ends of the bones. From these
growth centers, the bone continues to grow through
childhood and adolescence, finally ceasing in the early
twenties, whereupon the growth regions harden.

Diaphysis
The *diaphysis* (from Greek, meaning "a growing between")
is the shaft or central part of a long bone, and has a marrow-
filled cavity (medullary cavity) surrounded by compact bone.
It is formed from one or more primary sites of ossification,
and supplied by one or more nutrient arteries.

Epiphysis
The *epiphysis* (from Greek, meaning "excrescence") is the
end of a long bone, or any part of a bone separated from the
main body of an immature bone by cartilage. It is formed
from a secondary site of ossification, and consists largely
of spongy bone.

Epiphyseal Line
The *epiphyseal line* is the remnant of the epiphyseal plate
(a flat plate of hyaline cartilage) seen in young, growing
bone. It is the site of growth of a long bone. By the end of
puberty, long bone growth stops and this plate is completely
replaced by bone, leaving just the line to mark its previous
location.

Articular Cartilage
Articular cartilage is the only remaining evidence of an adult
bone's cartilaginous past, and is located where two bones
meet (articulate) within a synovial joint. It is smooth, slippery,

porous, malleable, insensitive, and bloodless. This type
of cartilage is massaged by movement, which permits the
absorption of synovial fluid, oxygen, and nutrition.

Note: The degenerative process of osteoarthritis, and the latter
stages of some forms of rheumatoid arthritis, involve the
breakdown of articular cartilage.

Periosteum
The *periosteum* is a fibrous connective tissue membrane that
is vascular and provides a highly sensitive double-layered
life-support sheath enveloping the outer surface of bone.
The outer layer is made of dense, irregular connective tissue.
The inner layer, which lies directly against the bone surface,
mostly comprises the bone-forming osteoblasts and the bone-
destroying osteoclasts.

The periosteum is supplied with nerve fibers, lymphatic
vessels, and blood vessels that enter the bone through nutrient
canals. It is attached to the bone by collagen fibers, known
as *Sharpey's fibers*, and also provides the anchoring point for
tendons and ligaments.

Medullary Cavity
The *medullary cavity* is the cavity of the diaphysis (i.e. the
central section of a long bone). It contains marrow: red in the
young, turning to yellow in many bones in maturity.

Red Marrow
Red marrow is a red, gelatinous substance composed of red
and white blood cells in a variety of developmental forms.
The red marrow cavities are typically found within the
spongy bone of long bones and flat bones. In adults the red
marrow, which creates new red blood cells, occurs only in
the head of the femur and the head of the humerus, and,

much more importantly, in the flat bones such as the sternum and the irregular bones such as the hip bones. These are the sites routinely used for obtaining red marrow samples when problems with the blood-forming tissues are suspected.

Yellow Marrow
Yellow marrow is a fatty connective tissue that no longer produces blood cells.

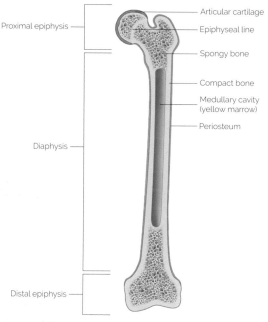

Proximal epiphysis —

Diaphysis —

Distal epiphysis —

Articular cartilage
Epiphyseal line
Spongy bone
Compact bone
Medullary cavity (yellow marrow)
Periosteum

Figure 3.8: Components of a long bone.

Bone Markings

Bone markings fall into three broad categories, as given below.

1. Projections on Bones That Are the Sites of Muscle and Ligament Attachments

Trochanter
A *trochanter* is very large, blunt, and irregularly shaped projection. The only example is on the femur.

Tuberosity
A *tuberosity* is a large rounded projection, which may be roughened. The main examples are on the tibia (tibial tuberosities) and the ischium (ischial tuberosities).

Tubercle
A *tubercle* is a smaller rounded projection, which may be roughened.

Crest
A *crest* is a projection, or projecting narrow ridge, of bone. It is usually prominent, a notable example being the iliac crest.

Border
A *border* is a bounding line or edge, also called a *margin*, or a narrow ridge of bone that separates two surfaces.

Spine or Spinous Process
A *spine* or *spinous process* is a sharp, slender, often pointed projection. Notable examples are the spinous processes on the vertebrae, and the spines of the scapula or the ilium (anterior superior iliac spine, abbreviated ASIS, and the posterior superior iliac spine, abbreviated PSIS).

Epicondyle
An *epicondyle* is a raised area, on or above a condyle, notably found on the humerus at the elbow joint.

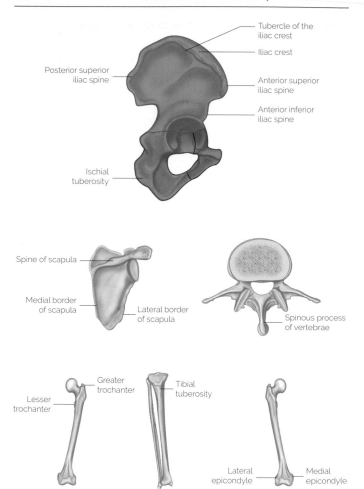

Figure 3.9: Projections on bones that are the sites of muscle and ligament attachments.

2. Projections on Bones That Help to Form Joints

Head

A *head* is an expansion, which is usually round, located at one end of a bone. An example is the head of the fibula, which articulates with the tibia, just below the knee joint.

Facet

A *facet* is a smooth, nearly flat surface at one end of bone, which articulates with another bone.

Condyle

A *condyle* is a large rounded projection, which articulates with another bone (as found at the knee joint).

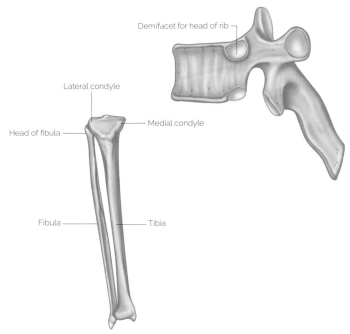

Figure 3.10: Projections on bones that help to form joints.

3. Depressions and Openings That Allow Blood Vessels and Nerves to Pass Through

Sinus

A *sinus* is a cavity within a bone, which is filled with air and lined with a membrane (most notably in the skull).

Fossa

A *fossa* is a shallow, basin-like depression in a bone, often serving as an articular surface.

Foramen

A *foramen* (plural = *foramina*) is a round or oval opening through a bone (most notably on the sacrum).

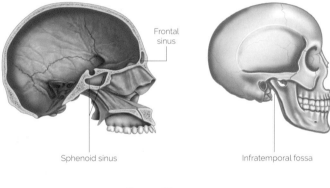

Frontal sinus

Sphenoid sinus

Infratemporal fossa

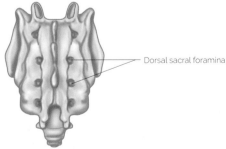

Dorsal sacral foramina

Figure 3.11: Depressions and openings that allow blood vessels and nerves to pass through.

CHAPTER 4

The Axial Skeleton

The Skull, Comprising the Cranium and the Facial Bones

The skull consists of 22 bones, excluding the ossicles of the ear. Except for the mandible, which forms the lower jaw, the bones of the skull are attached to each other by sutures, are immobile, and form the cranium (from Greek *kranion* = "the upper part of the head"). The cranium can be subdivided into the upper domed part (the calvaria); a base, consisting of the floor of the cranial cavity; and the lower anterior part (facial bones).

The Calvaria
The *calvaria* is made up of eight large flat bones, comprising two pairs of bones: temporal and parietal, and four single bones: frontal, occipital, sphenoid, and ethmoid. These bones form a box-like container that houses the brain.

Parietal
The *parietal bones* form most of the superior and lateral walls of the cranium. They meet in the midline at the sagittal suture, and meet with the frontal bone at the coronal suture.

Temporal
The *temporal bones* lie inferior to the parietal bones. There are three important markings on the temporal bone: (1) the *styloid process*, which is just in front of the mastoid process, a sharp

needle-like projection to which many of the neck muscles attach; (2) the *zygomatic process*, a thin bridge of bone that joins with the zygomatic bone, just above the mandible; and (3) the *mastoid process*, a rough projection posterior and inferior to the styloid process (just behind the lobe of the ear).

Frontal

The *frontal bone* forms the forehead, the bony projections under the eyebrows, and the superior part of each eye orbit.

Occipital

The *occipital bone* is the most posterior bone of the cranium. It forms the floor and back wall of the skull, and joins the parietal bones anteriorly at the lambdoid suture. In the base of the occipital bone is a large opening, the *foramen magnum*, through which the spinal cord passes to connect with the brain. To each side of the foramen magnum are the *occipital condyles*, which rest on the first vertebra of the spinal column (the *atlas*).

Sphenoid

The *sphenoid bone* is a butterfly-shaped bone that spans the width of the skull and forms part of the floor of the cranial cavity. Parts of the sphenoid can be seen forming part of the eye orbits, and the lateral part of the skull.

Ethmoid

The *ethmoid bone* is a single bone in front of the sphenoid bone and below the frontal bone. It forms part of the nasal septum and superior and medial conchae.

The Facial Bones

The face is composed of 13 bones, 12 of which are pairs. The paired bones of the face are: nasal, palatine, zygomatic, lacrimal, maxillary, and the inferior nasal concha. The unpaired bone is the vomer.

Nasal

The *nasal bones* are small rectangular bones that form the bridge of the nose (the lower part of the nose is made up of cartilage).

Palatine

The L-shaped palatine bones are situated at the back of the nasal cavity between the maxilla and the pterygoid process of the sphenoid bone.

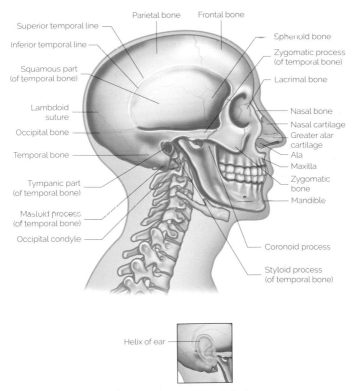

Figure 4.1: The skull (lateral view).

Zygomatic

The *zygomatic bones* are more commonly known as the *cheekbones*. They also form a large portion of the lateral walls of the eye orbits.

Lacrimal

The smallest of the facial bones, the lacrimal bones sit inside the bony orbit, and have two surfaces and four borders.

Maxillary

The two *maxillary bones*, or *maxillae*, fuse to form the upper jaw. The upper teeth are embedded in these bones.

Inferior Nasal Concha

These bones (pl. conchae) sit on the nasal septum, separating the nasal cavity into two bilateral and symmetrical anatomical caves.

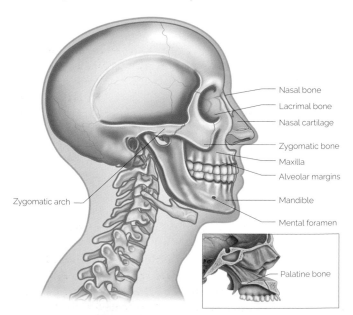

Figure 4.2: Facial bones (lateral view).

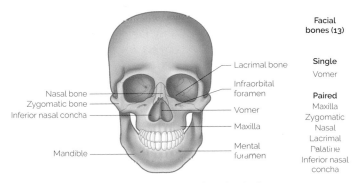

Figure 4.3: Facial bones (anterior view).

Vomer

The only single facial bone, the vomer runs vertically within the nasal cavity, separating the left and right sides. It is part of the nasal septum.

Mandible

Although neither categorised as a cranial or facial bone, the *mandible*, or *lower jawbone*, is the strongest bone in the skull; it joins the temporal bones on each side of the face, forming the only freely movable joints in the skull. The horizontal part of the mandible, or the body, forms the chin. Two upright bars of bone, or *rami*, extend from the body to connect the mandible with the temporal bone. The lower teeth are embedded in the mandible.

The Vertebral Column (Spine)

The vertebral column consists of 33 vertebrae in total:

7 **cervical** vertebrae; 12 **thoracic** vertebrae—which also form joints with the 12 ribs; 5 **lumbar** vertebrae—the largest, weight-bearing vertebrae; the *sacrum*, which is a single bone represented by 5 fused **sacral** vertebrae (note that the holes,

or foramina, in the sacrum correspond to the original gaps between the vertebrae); and the triangular *coccyx*, which represents 3 to 4 fused **coxal** vertebrae.

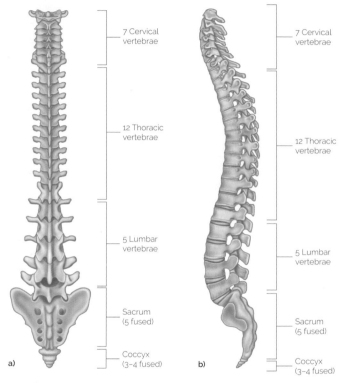

7 Cervical vertebrae

12 Thoracic vertebrae

5 Lumbar vertebrae

Sacrum (5 fused)

Coccyx (3–4 fused)

a)

7 Cervical vertebrae

12 Thoracic vertebrae

5 Lumbar vertebrae

Sacrum (5 fused)

Coccyx (3–4 fused)

b)

Figure 4.4: The vertebral column: (a) posterior view; (b) lateral view.

Typical Vertebra

The parts of a typical vertebra are: vertebral body, vertebral arch, vertebral foramen, transverse process, spinous process, and the superior and inferior articular processes.

The vertebral *body* is the disc-like, weight-bearing part of the vertebra. It faces anteriorly in the vertebral column. The size of the vertebral bodies increases inferiorly as the amount of weight being supported increases. So, for example, a lumbar vertebra is much larger than a cervical vertebra.

The *vertebral arch* forms the lateral and posterior parts of the vertebral foramen, and consists of pedicles and laminae. The two pedicles attach the vertebral arch to the vertebral body, and the two laminae extend from each pedicle to form the roof of the vertebral arch.

The *vertebral foramina* of all the vertebrae together form the vertebral canal, through which the spinal cord passes.

The *transverse process* extends posterolaterally from the junction of the pedicle and lamina on each side and acts as a site for articulation with the ribs.

The *spinous process* is a single projection that arises from the posterior part of the vertebral arch. On the cervical vertebrae, the spinous processes are short and divide into two points (this division looks a little like a whale's tail). On the thoracic vertebrae, the spinous processes are single, slender points that angle sharply downward. On the lumbar vertebrae, the spinous processes are thick and wedge shaped.

The *superior/inferior articular processes* are paired projections lateral to the foramen. They allow one vertebra to form a joint with the next vertebra.

The illustrations on the following pages give a selection of key vertebrae shown from various angles, in order to highlight the variations in shape and features.

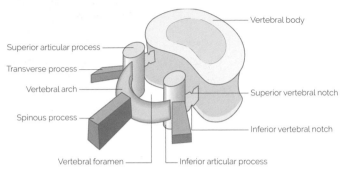

Figure 4.5: Schematic of a typical vertebra.

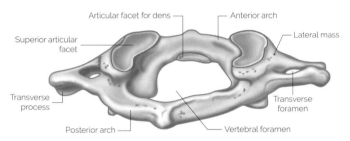

Figure 4.6: Atlas (C1) posterosuperior view.

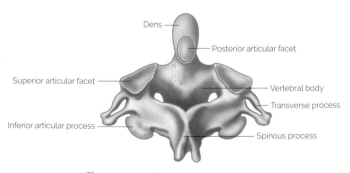

Figure 4.7: Axis (C2) posterosuperior view.

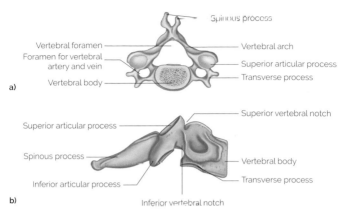

Figure 4.8: (a) Cervical vertebra (C5) superior view.
(b) Cervical vertebra (C5) lateral view.

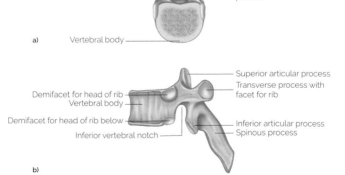

Figure 4.9: (a) Thoracic vertebra (T6) superior view.
(b) Thoracic vertebra (T6) lateral view.

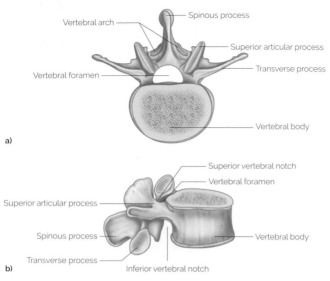

Figure 4.10: (a) Lumbar vertebra (L3) superior view.
(b) Lumbar vertebra (L3) lateral view.

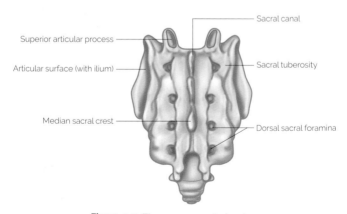

Figure 4.11: The sacrum: posterior view.

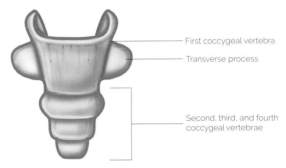

First coccygeal vertebra

Transverse process

Second, third, and fourth coccygeal vertebrae

Figure 4.12: The coccyx: posterior view.

The Bony Thorax

The Sternum

The *sternum* is commonly known as the *breastbone*. It is actually the fusion of three bones: the manubrium, the body (also known as the *gladiolus*), and the xiphoid process.

Note: The sternum is attached to the first seven pairs of ribs by the costal cartilage. *Manubrium* means "handle," as in the handle of a sword; *xiphoid* means "sword shaped."

The Ribs

The *ribs* consist of 12 pairs in total, and comprise true, false, and floating ribs:

- *True ribs*: the first seven pairs attach by costal cartilage directly to the sternum.
- *False ribs*: the next three pairs attach to costal cartilage, but not directly to the sternum.
- *Floating ribs*: the last two pairs of ribs lack attachments either to costal cartilage or to the sternum.

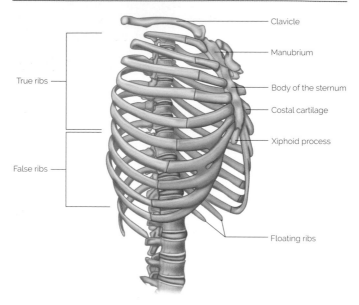

True ribs

False ribs

Clavicle

Manubrium

Body of the sternum

Costal cartilage

Xiphoid process

Floating ribs

Figure 4.13: The bony thorax.

CHAPTER 5

The Appendicular Skeleton

The Pectoral Girdle

Note: Some soft tissues (e.g. cartilages, aponeuroses, ligaments, and tendons) are included where appropriate, for ease of reference.

Clavicle

The *clavicle* is commonly known as the *collarbone*. It is a slender, doubly curved bone that attaches to the manubrium of the sternum medially (the *sternoclavicular joint*) and to the acromion of the scapula laterally (the *acromioclavicular joint*).

Scapula

Commonly known as the *shoulder blade*, the *scapula* is a large triangular flat bone lying posterior to the dorsal thorax, between the second and the seventh ribs. Each scapula articulates with the clavicle and the humerus, and has four important bone markings:

1. *Spine*: a sharp, prominent ridge on the posterior surface of the scapula, which can be easily felt through the skin.
2. *Acromion*: an enlarged anterior projection at the lateral end of the spine of the scapula, which can be felt as the "point of the shoulder."

3. *Coracoid process*: a forward projection from the upper border of the scapula.
4. *Glenoid fossa*: a shallow depression at the lateral angle of the scapula, which articulates with the head of the humerus.

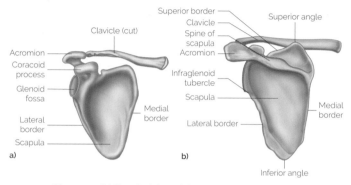

Figure 5.1: (a) The clavicle and the scapula (anterior view). (b) The scapula (posterior view).

The Upper Limb

Humerus

The *humerus* (or *arm bone*) is the longest and largest bone of the upper limb. It articulates proximally with the scapula (at the glenoid fossa). At the distal end is the *trochlea* (which looks like a spool) and the *capitulum* (or head), which together form part of the elbow joint with the ulna and the radius. On either side of the trochlea are the medial and lateral epicondyles of the humerus, easily felt superficially.

Radius

The *radius* is one of the two bones in the forearm, lying on the lateral, or thumb side, of the forearm. Proximally, the head of the radius forms a joint with the capitulum of the humerus. The radius crosses the ulna during pronation.

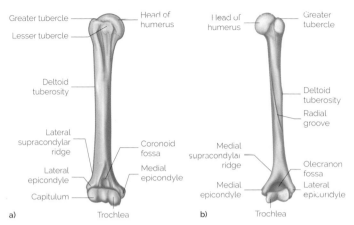

Greater tubercle

Lesser tubercle

Head of humerus

Head of humerus

Greater tubercle

Deltoid tuberosity

Deltoid tuberosity

Radial groove

Lateral supracondylar ridge

Coronoid fossa

Medial supracondylar ridge

Olecranon fossa

Lateral epicondyle

Medial epicondyle

Medial epicondyle

Lateral epicondyle

Capitulum

a)

Trochlea

b)

Trochlea

Figure 5.2: (a) The right humerus (anterior view).
(b) The right humerus (posterior view).

Ulna

The *ulna* is the medial bone in the forearm, on the little-finger side. At the proximal end of the ulna are two processes: the coronoid and the olecranon, which fit over the two medial rounded spools of the trochlea of the humerus. The *coronoid process* is a projection from the anterior portion of the proximal end of the ulna, forming part of the articulation of the elbow. The *olecranon process* is the pointed bump felt when the elbow is bent, and is also known as the *funny bone*, because when the nerve that runs over the olecranon is hit, it can be painful. The styloid processes of the radius and ulna can be felt as sharp projections on either side of the wrist.

8 Carpals

The *carpals* are eight small bones that make up the wrist. They are bound together by ligaments and are arranged in two transverse rows, four bones to a row. The first row comprises the scaphoid, lunate, triquetrum, and pisiform. The second row comprises the trapezium, trapezoid, capitate, and hamate. A mnemonic for memorizing the carpals from lateral to medial, beginning with the proximal row, is: "some lovers try positions that they cannot handle."

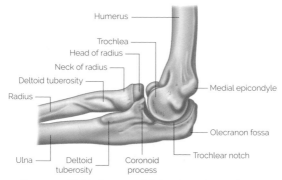

Humerus

Trochlea

Head of radius

Neck of radius

Deltoid tuberosity

Radius

Medial epicondyle

Olecranon fossa

Trochlear notch

Ulna

Deltoid tuberosity

Coronoid process

Figure 5.3: Right elbow: medial view in 90-degree flexion.

5 Metacarpals

The *metacarpals* are five bones running between the wrist and the knuckles (which are the heads of the metacarpals).

14 Phalanges

The *phalanges* (sing. *phalanx*) are the bones forming the distal segment of the skeleton of each limb. Each finger has three phalanges, whereas the thumb has only two.

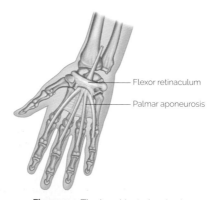

Flexor retinaculum

Palmar aponeurosis

Figure 5.4: The hand (anterior view).

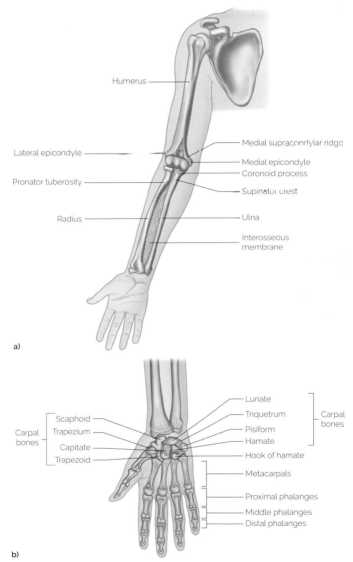

Figure 5.5: The bones of the right (a) forearm and (b) hand (anterior view).

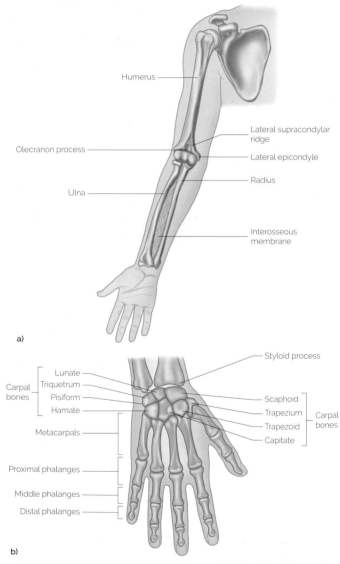

Figure 5.6: The bones of the right (a) forearm and (b) hand (posterior view).

The Pelvic Girdle (Os Innominatum)

The *pelvic girdle* (or *hip girdle*) consists of two pelvic, or coxal, bones. It provides a strong and stable support for the lower extremities on which the weight of the body is carried. The pelvic bones unite with one another in the front (anteriorly) at the *pubic symphysis* (a fibrocartilaginous disc). With the sacrum and coccyx, the two pelvic bones form a basin-like structure called the *pelvis*. At birth, each pelvic bone consists of three separate bones: the ilium, the ischium, and the pubis. These three bones eventually fuse into one pelvic bone, and the area where they join is a deep hemispherical socket called the *acetabulum* (this socket articulates with the head of the femur). Although the pelvic bone is a single bone, it is still commonly discussed as if it consisted of three portions.

Ilium
The *ilium* is a large, flaring bone that forms the largest and most superior portion of the pelvic bone. The iliac crests can be felt when you rest your hands on your hips. Each crest terminates at the front as the ASIS, and at the back as the PSIS (the PSIS is difficult to palpate, but its position is revealed by a skin dimple in the sacral region, approximately level with the second sacral foramen).

Ischium
The *ischium* is the inferior, posterior part of the pelvic bone, and is roughly arch shaped. At the bottom of the ischium are the roughened and thickened ischial tuberosities (sometimes called the *sit bones*, because when we sit, our weight is borne entirely by the ischial tuberosities).

Pubis
The *pubis* is the anterior and inferior part of the pelvic bone.

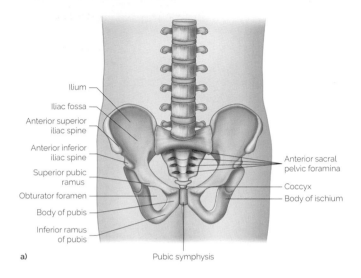

Ilium
Iliac fossa
Anterior superior iliac spine
Anterior inferior iliac spine
Superior pubic ramus
Obturator foramen
Body of pubis
Inferior ramus of pubis

Anterior sacral pelvic foramina
Coccyx
Body of ischium

a)

Pubic symphysis

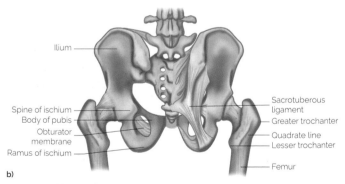

Ilium

Spine of ischium
Body of pubis
Obturator membrane
Ramus of ischium

Sacrotuberous ligament
Greater trochanter
Quadrate line
Lesser trochanter

Femur

b)

Figure 5.7: (a) Bones of the pelvic girdle (anterior view).
(b) Bones of the pelvic girdle (posterior view).

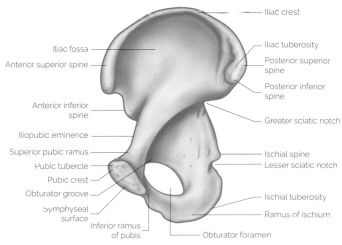

Figure 5.8: Right pelvic (coxal) bone (posterior view).

The Lower Limb

Femur

The *femur* is the only bone in the thigh, and is the heaviest, longest, and strongest bone in the body. Its proximal end has a ball-like head that articulates with the pelvic bone at the acetabulum. Distally on the femur are the lateral and medial condyles, which articulate with the tibia.

The *greater trochanter* is a projection just distal to the head and neck of the femur and can sometimes be felt in the buttock.

Tibia

The *tibia* (or *shin bone*) is the larger and more medial of the bones in the lower leg. At the proximal end, the medial and lateral condyles articulate with the distal end of the femur to form the knee joint.

The *tibial tuberosity* is a roughened area on the anterior surface of the tibia. The *medial malleolus* can be felt as the inner bulge of the ankle.

Fibula

The *fibula* lies lateral and parallel to the tibia, and is thin and sticklike. It is not a weight-bearing bone, and moreover plays no part in the knee joint.

The *lateral malleolus* on the fibula can be felt as the outer bulge of the ankle.

Patella

Known as the *kneecap*, the *patella* is a small triangular sesamoid bone within the tendon of the quadriceps femoris muscle. It forms the front of the knee joint.

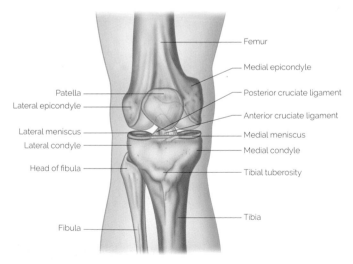

Figure 5.9: Lower femur and upper tibia and upper fibula of the right leg (anterior view).

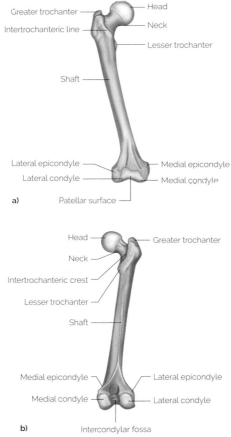

Figure 5.10: Femur of the right leg: (a) anterior view; (b) posterior view.

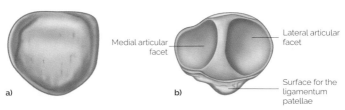

Figure 5.11: Patella of the right leg: (a) anterior view; (b) posterior view.

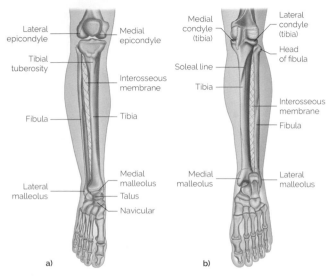

Figure 5.12: Tibia and fibula of the right leg: (a) anterior view; (b) posterior view.

7 Tarsals

The *tarsals* are the seven bones of the ankle. The two largest tarsals, mainly carrying the body weight, are the calcaneus (or the heel bone) and the talus, which lies between the tibia and the calcaneus. The navicular, medial cuneiform, intermediate cuneiform, lateral cuneiform, and cuboid constitute the other five tarsals.

5 Metatarsals

The *metatarsals* form the instep or sole of the foot.

14 Phalanges

The *phalanges* are the bones forming the distal segment of the skeleton of each limb. Each toe has three phalanges, except the big toe, which has only two.

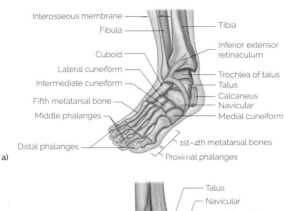

Interosseous membrane
Fibula
Cuboid
Lateral cuneiform
Intermediate cuneiform
Fifth metatarsal bone
Middle phalanges
Distal phalanges
Tibia
Inferior extensor retinaculum
Trochlea of talus
Talus
Calcaneus
Navicular
Medial cuneiform
1st–4th metatarsal bones
Proximal phalanges

a)

Talus
Navicular
Cuneiform
Calcaneus
Cuboid
Metatarsals
Phalanges

b)

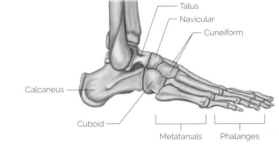

Medial malleolus of tibia
Talus
Tuberosity of navicular
Intermediate cuneiform
Medial cuneiform
Lateral cuneiform
Metatarsal bones
Proximal phalanges
Tuberosity of calcaneus
Medial border of calcaneus
Sustentaculum tali
Lateral border of calcaneus
Cuboid
Middle phalanges
Distal phalanges

c)

Figure 5.13: Bones of the right foot: (a) anteromedial view;
(b) lateral view; (c) plantar view.

General Skeletal Interrelationships

Note: Some soft tissues (e.g. cartilages, aponeuroses, ligaments, and tendons) are included where appropriate, for ease of reference.

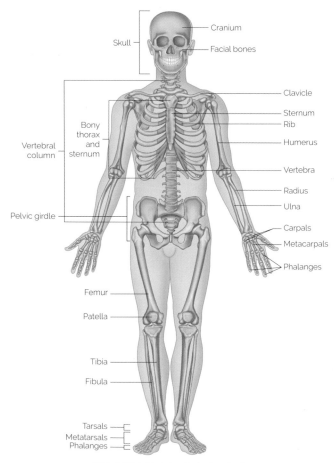

Figure 5.14: Skeleton (anterior view).

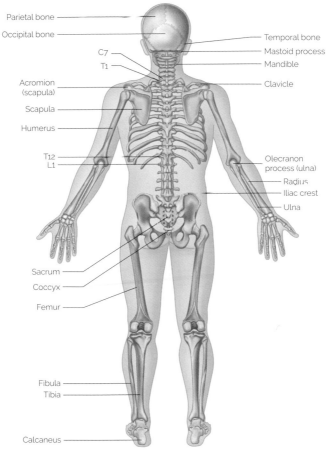

Figure 5.15: Skeleton (posterior view).

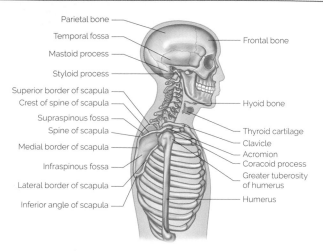

Figure 5.16: Skull to humerus (lateral view).

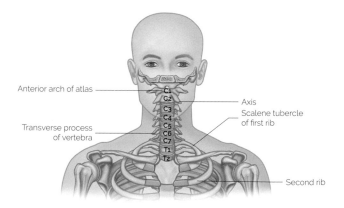

Figure 5.17: Skull to sternum (anterior view, the mandible and maxilla are removed).

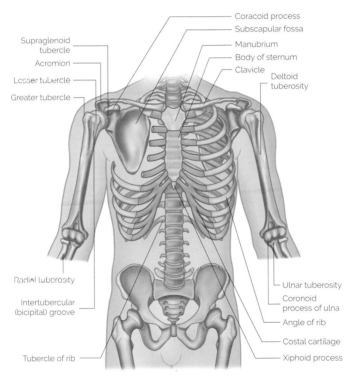

Coracoid process
Subscapular fossa
Manubrium
Body of sternum
Clavicle
Deltoid tuberosity

Supraglenoid tubercle
Acromion
Lesser tubercle
Greater tubercle

Radial tuberosity
Intertubercular (bicipital) groove
Tubercle of rib

Ulnar tuberosity
Coronoid process of ulna
Angle of rib
Costal cartilage
Xiphoid process

Figure 5.18: Ribcage, pectoral girdle, upper arm (anterior view, the upper right anterior ribcage is removed).

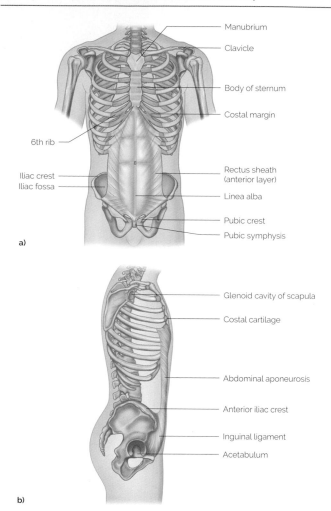

Manubrium

Clavicle

Body of sternum

Costal margin

6th rib

Iliac crest

Iliac fossa

Rectus sheath
(anterior layer)

Linea alba

Pubic crest

Pubic symphysis

a)

Glenoid cavity of scapula

Costal cartilage

Abdominal aponeurosis

Anterior iliac crest

Inguinal ligament

Acetabulum

b)

Figure 5.19: (a) Thoracic to pelvic region (anterior view).
(b) Thoracic to pelvic region (lateral view).

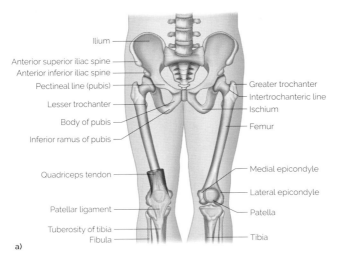

Ilium

Anterior superior iliac spine

Anterior inferior iliac spine

Pectineal line (pubis)

Lesser trochanter

Body of pubis

Inferior ramus of pubis

Quadriceps tendon

Patellar ligament

Tuberosity of tibia

Fibula

a)

Greater trochanter

Intertrochanteric line

Ischium

Femur

Medial epicondyle

Lateral epicondyle

Patella

Tibia

Body of pubis

Inferior ramus of pubis

Ischial tuberosity

Ramus of ischium

Lateral supracondylar line

Adductor tubercle

Lateral condyle of femur

Lateral condyle of tibia

Head of fibula

Fibula

b)

Sacrotuberous ligament

Greater trochanter

Lesser trochanter

Gluteal tuberosity

Linea aspera

Femur

Medial supracondylar line

Popliteal surface

Medial condyle of femur

Medial condyle of tibia

Tibia

Figure 5.20: (a) Pelvic girdle to leg (anterior view).
(b) Pelvic girdle to leg (posterior view).

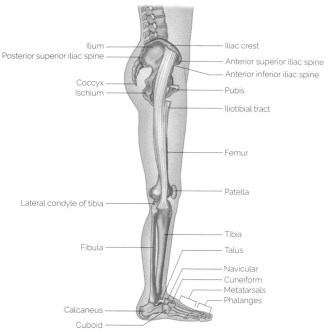

Ilium
Posterior superior iliac spine
Iliac crest
Anterior superior iliac spine
Anterior inferior iliac spine
Coccyx
Ischium
Pubis
Iliotibial tract
Femur
Patella
Lateral condyle of tibia
Tibia
Fibula
Talus
Navicular
Cuneiform
Metatarsals
Phalanges
Calcaneus
Cuboid

Figure 5.21: Pelvic girdle to foot (lateral view).

Bony Landmarks Seen or Felt Near the Body Surface

The following bony landmarks can be seen or felt near the surface of the body. Identify them on yourself or a partner, using Figure 5.22(a–c) for reference.

Frontal bone
Temporal bone
Occipital bone
Manubriosternum and manubriosternal joint (level with the 2nd rib)

2nd rib
Sternoclavicular joint
Acromioclavicular joint
Spine of the scapula
Medial border of the scapula
Inferior angle of the scapula
Medial and lateral epicondyles of the humerus
Olecranon
Head of the radius
Ulnar styloid
Pisiform bone
Anatomical snuffbox
Iliac crest
Anterior superior iliac spine (ASIS)
Posterior superior iliac spine (PSIS)
Ischial tuberosities
Greater trochanter
Head of the fibula
Tibial tuberosity
Medial and lateral malleoli
Calcaneus
Spinous processes of the vertebrae

Hints
C2:   the first cervical vertebra to be felt below the occiput.
C7:   at the base of the neck, the vertebra that stands out most
      prominently.
T3–4: level with the spine of the scapula.
T7:   level with the inferior angle of the scapula.
L4:   level with the iliac crest.
S2:   level with the PSIS (or visible as the dimple at the top of
      the buttocks).

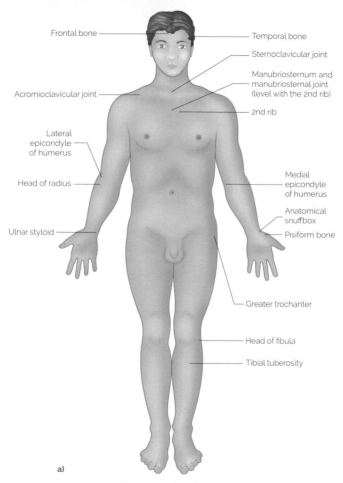

a)

Figure 5.22: (continued)

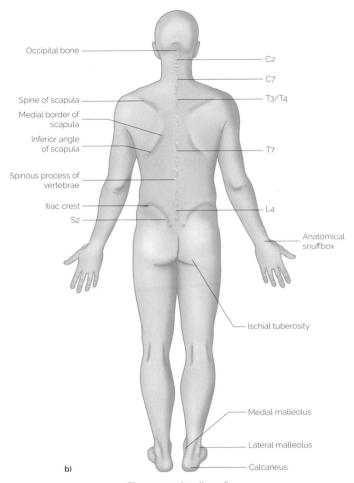

Occipital bone

C2

C7

Spine of scapula

T3/T4

Medial border of scapula

Inferior angle of scapula

T7

Spinous process of vertebrae

Iliac crest

L4

S2

Anatomical snuffbox

Ischial tuberosity

Medial malleolus

Lateral malleolus

b)

Calcaneus

Figure 5.22: (continued)

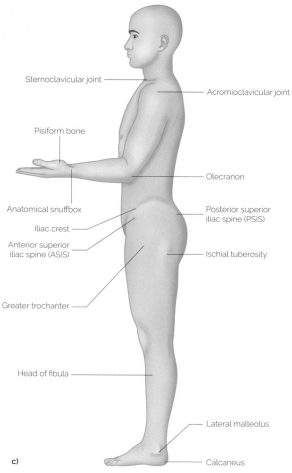

Sternoclavicular joint

Acromioclavicular joint

Pisiform bone

Olecranon

Anatomical snuffbox

Posterior superior iliac spine (PSIS)

Iliac crest

Anterior superior iliac spine (ASIS)

Ischial tuberosity

Greater trochanter

Head of fibula

Lateral malleolus

c)

Calcaneus

Figure 5.22: (a–c) Bony landmarks.

CHAPTER 6

Joints

With the exception of the hyoid bone in the neck, all bones form a joint with at least one other bone. Joints are also called *articulations*.

Joints have two functions: (1) to hold the bones together; and (2) to give the rigid skeleton mobility. When two bones meet, or articulate, there may or may not be movement, depending on: (1) the amount of bonding material between the bones; (2) the nature of the material between the bones; (3) the shape of the bony surfaces; (4) the amount of tension in the ligaments or muscles involved at the joint; and (5) the position of the ligaments and muscles.

Part One—Classification of Joints

Joints are classified in two ways: by function and by structure.

Function
The functional classification of joints focuses on the amount of movement allowed by the joint.

Immovable Joints (Synarthrotic)
Immovable joints are found mostly in the axial skeleton, where joint stability and firmness is important for the protection of the internal organs.

Slightly Movable Joints (Amphiarthrotic)

Like immovable joints, and for the same reason, *slightly movable joints* are also found mainly in the axial skeleton.

Freely Movable Joints (Diarthrotic)

Freely movable joints predominate in the limbs, where a greater range of movement is required.

Structure

Fibrous Joints

In *fibrous joints*, fibrous tissue joins the bones; as a result, no joint cavity is present. Generally these joints have little or no movement, i.e. they are synarthrotic. There are three types of fibrous joint: suture, syndesmosis, and gomphosis.

The only examples of *fibrous sutures* are the sutures of the skull, where the irregular edges of the bones interlock and are bound tightly together by connective tissue fibers, allowing no active movement. Layers of periosteum on the inner and outer layers of the adjoining bones bridge the gap between the bones and form the main bonding factor. Between the adjoining joint surfaces there is a layer of vascular fibrous tissue that also helps unite the bones. This vascular fibrous tissue, plus the two layers of periosteum, are collectively called the *sutural ligament*. The fibrous tissue becomes ossified during adulthood by a process that occurs first at the deep aspect of the suture, progressively extending to the superficial part. This ossifying process is referred to as *synostosis*.

A *syndesmosis* is a fibrous joint where the uniting fibrous tissue forms an interosseous membrane or ligament, i.e. a band of fibrous tissue that allows little movement. Examples are the interosseous membranes situated between the radius and the ulna and between the tibia and the fibula.

A *gomphosis* refers to a fibrous joint in which a peg is embedded into a socket. The only examples of such joints in humans are the teeth fixed into their sockets.

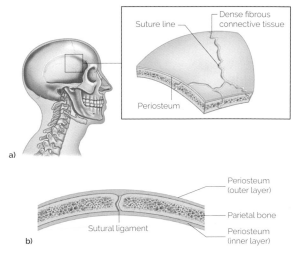

a)

b)

Figure 6.1: (a) Position of a suture; (b) vertical section through a suture.

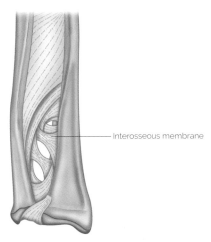

Figure 6.2: The interosseous membrane between the radius and ulna.

Cartilaginous Joints

In *cartilaginous joints*, a continuous plate of hyaline cartilage or a fibrocartilaginous disc connects the bones; again, no joint cavity is present. These joints can be either immovable (synchondroses) or slightly movable (symphyses), with the slightly movable joints being the more common.

Examples of cartilaginous joints that are immovable are the epiphyseal plates of growing long bones. These plates are made of hyaline cartilage, which ossifies in young adults (see p. 33). The place where a joint is united by such a plate is known as a *synchondrosis*. Another example of such a joint that eventually ossifies is the joint between the first rib and the manubrium of the sternum.

Examples of slightly movable cartilaginous joints are the pubic symphysis of the pelvic girdle, and the intervertebral joints of the spinal column. In both cases the articular surfaces of the bones are covered with hyaline cartilage, which is in turn fused to a "pad" of fibrocartilage (fibrocartilage is compressible and resilient, and acts as a shock absorber).

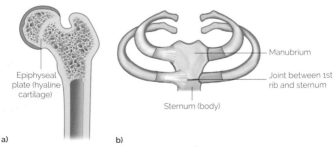

Epiphyseal plate (hyaline cartilage)

Manubrium

Joint between 1st rib and sternum

Sternum (body)

a)   b)

Figure 6.3: Cartilaginous immovable (synchondrosis) joints (anterior view): (a) the epiphyseal plate in a growing long bone; (b) the sternocostal joint between the manubrium and the first rib.

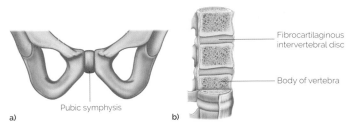

Figure 6.4: Cartilaginous slightly movable (amphiarthrotic/symphysis) joints (anterior view): (a) the pubic symphysis of the pelvic girdle, (b) the intervertebral joints.

Synovial Joints

Synovial joints possess a joint cavity that contains synovial fluid. They are freely movable (diarthrotic) joints, and have a number of distinguishing features.

Articular cartilage (or *hyaline cartilage*) covers the ends of the bones that form the joint.

The *joint cavity* is more a potential space than a real one, because it is filled with lubricating synovial fluid. The joint cavity is enclosed by a double-layered "sleeve," or capsule, known as the *articular capsule*.

The external layer of the articular capsule is known as the *capsular ligament*. This is a tough, flexible, fibrous connective tissue that is continuous with the periostea of the articulating bones. The internal layer, or *synovial membrane*, is a smooth membrane made of loose connective tissue that lines the capsule and all internal joint surfaces other than those covered in hyaline cartilage.

Synovial fluid is a slippery fluid that occupies the free spaces within the joint capsule; it is also found within the articular cartilage and provides a film that reduces friction between the cartilages. When a joint is compressed by movement, synovial fluid is forced out of the cartilage; when the pressure is relieved, the fluid rushes back into the articular cartilage. Synovial fluid nourishes the cartilage, which is avascular (contains no blood vessels); it also contains phagocytic cells (cells that eat dead matter), which rid the joint cavity of microbes or cellular waste. The amount of synovial fluid varies in different joints, but is always sufficient for forming a thin film to reduce friction. During injury to the joint, extra fluid is produced and creates the characteristic swelling of the joint. This extra fluid is later reabsorbed by the synovial membrane.

Collateral or *accessory ligaments* reinforce and strengthen the synovial joints; these ligaments are either capsular (i.e. thickened parts of the fibrous capsule itself) or independent (i.e. distinct from the capsule). Ligaments always bind bone to bone, and, according to their position and quantity around the joint, they will restrict movement in certain directions, as well as preventing unwanted movement. As a general rule, the more ligaments a joint has, the stronger it is.

Bursae (sing. *bursa*) are fluid-filled sacs often found cushioning the joint; they are lined by synovial membrane and contain synovial fluid. Bursae are found between tendon and bone, ligament and bone, and muscle and bone, and reduce friction by acting as a cushion.

Tendon sheaths are also frequently found in close proximity to synovial joints. They have the same structure as bursae, and wrap themselves around tendons subject to friction, in order to protect them.

Articular discs (*menisci*) are present in some synovial joints. They act as shock absorbers (similar to the fibrocartilaginous disc in the pubic symphysis). For example, in the knee joint, two crescent-shaped fibrocartilaginous discs called the *medial meniscus* and the *lateral meniscus* lie between the medial condyles of the femur and tibia and between the lateral condyles of the same two bones.

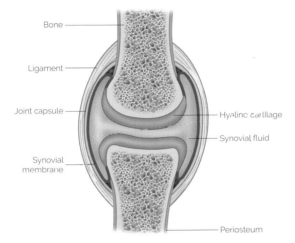

Figure 6.5: A typical synovial joint.

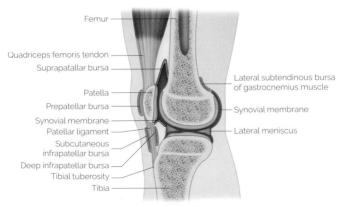

Figure 6.6: Shock-absorbing and friction-reducing structures of a synovial joint.

There are seven types of synovial joint: plane (or gliding), hinge, pivot, ball-and-socket, condyloid, saddle, and ellipsoid.

In *plane joints* (or *gliding joints*), movement occurs when two, generally flat or slightly curved, surfaces glide across one another. Examples: acromioclavicular joint, joints between the carpal bones in the wrist and between the tarsal bones in the ankle, facet joints between the vertebrae, sacroiliac joint.

In *hinge joints*, movement occurs around only one axis—a transverse one—as in the hinge of the lid of a box.
A protrusion of one bone fits into a concave or cylindrical articular surface of another, permitting flexion and extension. Examples: interphalangeal joints, elbow, knee.

In *pivot joints*, movement takes place around a vertical axis, like the hinge of a gate. A more or less cylindrical articular surface of bone protrudes into and rotates within a ring formed by bone or ligament. Examples: (1) the dens of the axis protrudes through the hole in the atlas, allowing the rotation of the head from side to side; (2) the joint between the radius and the ulna at the elbow allows the round head of the radius to rotate within a "ring" of ligament that is secured to the ulna.

Ball-and-socket joints consist of a "ball" formed by the spherical or hemispherical head of one bone, which rotates within the concave "socket" of another, allowing flexion, extension, adduction, abduction, circumduction, and rotation. Thus, they are multiaxial and allow the greatest range of movement of all joints. Examples: shoulder and hip joints.

In common with ball-and-socket joints, *condyloid joints* have a spherical articular surface that fits into a matching concavity. As in the case of ball-and-socket joints, condyloid joints also permit flexion, extension, abduction, adduction, and circumduction; however, the disposition of surrounding ligaments and muscles prevents active rotation around a vertical axis. Examples: metacarpophalangeal joints of the fingers (but not the thumb).

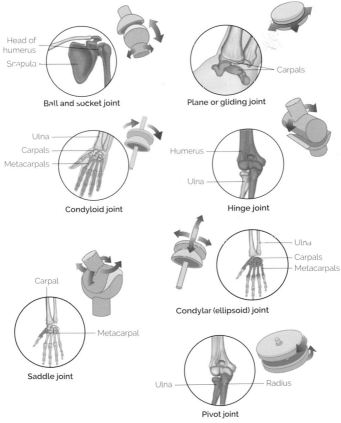

Figure 6.7: Types of synovial joint.

Saddle joints are similar to condyloid joints, except that both articulating surfaces have convex and concave areas, which fit together like a saddle and a horse's back. These joints allow even more movement than condyloid joints. Example: carpometacarpal joint of the thumb, which allows opposition of the thumb to the fingers.

An *ellipsoid joint* is effectively similar to a ball-and-socket joint, but the articular surfaces are ellipsoidal instead of spherical. Movements are the same as for ball-and-socket joints, with the exception of rotation (the shape of the ellipsoid surfaces prevents this). Example: radiocarpal joint.

Part Two—Features of Specific Joints

This section presents a relatively in-depth examination of four synovial joints: shoulder, elbow, hip, and knee. The other joints, of all classifications, are presented in less detail, with minimal accompanying text, but with fully labelled illustrations.

Notes About Synovial Joints
- Ligaments are termed *intracapsular* (or *intra-articular*) when inside the joint cavity, and *extracapsular* (or *extra-articular*) when outside the capsule.
- Some tendons run partly within the joint and are therefore intracapsular.
- The fibers of many ligaments are largely integrated with those of the capsule, and the delineation between capsule and ligament is sometimes unclear; therefore, only the main ligaments are mentioned.
- Many ligaments of the knee joint are modified extensions or expansions of muscle tendons, but are classed as *ligaments* to differentiate them from the more regular stabilizing tendons, such as the patellar ligament from the quadriceps.

- Most synovial joints have various bursae in their vicinity, as shown in the illustrations pertaining to each joint.

Joints of the Head and Vertebral Column

Temporomandibular Joint
Type of Joint
Synovial hinge joint, and plane joint.

Articulation
The head of the mandible articulates with the mandibular fossa and the articular tubercle of the temporal bone. A fibrous disc separates the articular surfaces and molds itself upon them when the joint moves.

Movements
This is the only movable joint in the head. Movement can occur in all three planes: upward and downward, backward and forward, and from side to side. A gliding action occurs superior to the disc, while a hinge action occurs inferior to the disc.

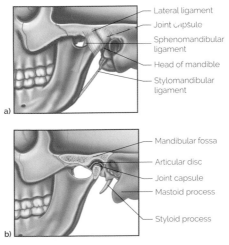

Lateral ligament
Joint capsule
Sphenomandibular ligament
Head of mandible
Stylomandibular ligament

a)

Mandibular fossa
Articular disc
Joint capsule
Mastoid process
Styloid process

b)

Figure 6.8: (a&b) The temporomandibular joint (lateral view).

Atlanto-occipital Joint
Type of Joint
The articulations of the two sides act together functionally as a synovial ellipsoid joint.

Articulation
Between the occipital condyles and the superior articular facets of the atlas.

Movements
Flexion and extension (as in nodding the head). Lateral flexion.

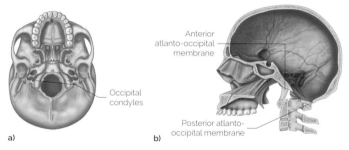

Anterior atlanto-occipital membrane

Occipital condyles

Posterior atlanto-occipital membrane

a)   b)

Figure 6.9: The atlanto-occipital joint: (a) inferior view; (b) lateral view.

Atlanto-axial Joint
Type of Joint
Lateral atlanto-axial joint: synovial plane.
Median (sagittal) atlanto-axial joint: synovial pivot.

Articulation
Lateral atlanto-axial joint: between the opposed articular processes of the atlas and the axis.

Median (sagittal) atlanto-axial joint: between the dens of the axis and the anterior arch of the atlas, and with the transverse ligament.

Movements

Rotation of the head around a vertical axis (the skull and the atlas moving as one).

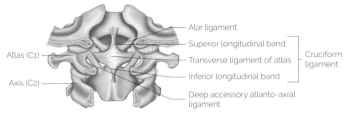

Figure 6.10: The atlanto-axial joint (posterior view).

Joints Between Vertebral Bodies

Type of Joint

Cartilaginous symphysis (slightly movable).

Articulation

Between the adjacent surfaces of the vertebral bodies, and united by a fibrocartilaginous intervertebral disc.

Movements

Only slight movement occurs between any two successive vertebrae, but there is considerable movement throughout the column as a whole.

Cervical region: flexion, extension, lateral flexion with rotation (i.e. lateral flexion cannot occur without an element of rotation, and vice versa).

Thoracic region: rotation, always associated with an element of lateral flexion, and vice versa; only extremely slight flexion and extension can occur (limited by the presence of the ribs and sternum).

Lumbar region: flexion, extension; only extremely slight rotation can occur (restricted by the angle of the articular processes).

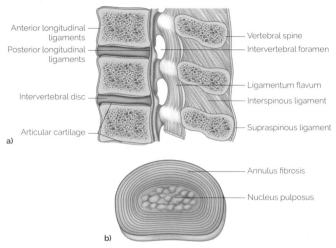

Figure 6.11: (a) Sagittal section through the 2nd to 4th lumbar vertebrae. (b) Transverse section of a lumbar intervertebral disc.

Joints Between Vertebral Arches
Type of Joint
Synovial plane.

Articulation
Between opposed articular processes of adjacent vertebrae, to unite adjacent vertebral arches.

Movements
Only slight movement occurs between any two successive vertebrae, but there is considerable movement throughout the column as a whole.

Cervical region: flexion, extension, lateral flexion with rotation (i.e. lateral flexion cannot occur without an element of rotation, and vice versa).

Thoracic region: rotation, always associated with an element of lateral flexion, and vice versa; only extremely slight flexion and extension can occur (limited by presence of ribs and sternum).

Lumbar region: flexion, extension; only extremely slight rotation can occur (restricted by the angle of the articular processes).

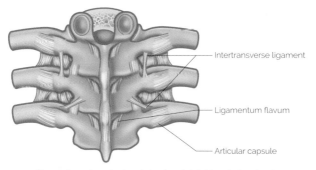

Intertransverse ligament

Ligamentum flavum

Articular capsule

Figure 6.12: A typical vertebral arch joint (posterior view).

Joints of the Ribs and Sternum

Costovertebral Joints
Type of Joint
Joints of the heads of the ribs (capitular joints): synovial plane.
Costotransverse joints: synovial plane.

Articulation
Joints of the heads of the ribs (capitular joints): superior and inferior articular facets on the head of each typical rib articulate with the facets on the two adjacent vertebral bodies (i.e. the rib's head sits between two vertebral bodies, and also against a shallow depression on the intervertebral disc).

Costotransverse joints: the tubercle of each typical rib articulates with the transverse process of the lower of the two vertebrae to which its head is joined (but ligaments attach it to the transverse processes of both vertebrae).

Note: The first rib and the last two or three ribs have atypical vertebral connections, because the heads of these ribs have only one facet, not two; they therefore articulate with one vertebral body rather than two. The tubercles of the lowest ribs do not form synovial joints with the transverse processes.

Movements

The capitular and costotransverse joints of each rib together form a hinge, causing the anterior part of the rib to be raised (with some lateral "expansion") during inspiration, and to be lowered (with some medial "contraction") during expiration. This effectively increases and decreases the anteroposterior and transverse diameters of the thorax with each in-breath and out-breath.

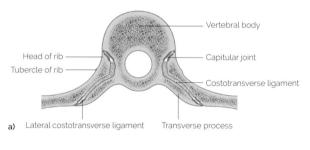

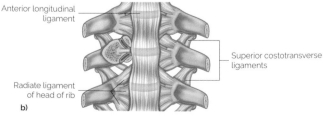

Figure 6.13: (a) Transverse section through a typical costovertebral joint. (b) The costovertebral joint (anterior view).

Sternocostal Joints
The hyaline cartilage that is continuous with the anterior end of each rib is called the *costal cartilage*.

Type of Joint
First rib: cartilaginous immovable (synchondrosis).
Ribs 2–7: simple synovial plane.
Ribs 8–10: simple synovial plane articulations at the interchondral joints.

Articulation
First rib: via costal cartilage to the body of the sternum.
Ribs 2–7: via costal cartilages to facets on the side of the body of the sternum; the joint cavities are divided into two by an intra-articular ligament (until cavities disappear in old age).
Ribs 8–10: their costal cartilages unite with the costal cartilage of rib 7.
Ribs 11–12: do not articulate anteriorly, but end freely in the muscles of the flank; they are therefore called *floating ribs*.

Movements
Enables expansion and contraction of the ribcage, as described under "Costovertebral Joints" (p. 97).

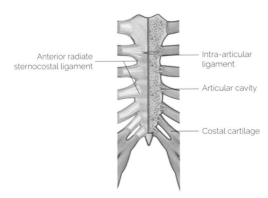

Anterior radiate sternocostal ligament

Intra-articular ligament

Articular cavity

Costal cartilage

Figure 6.14: The sternocostal joint (anterior view).

Sternal Joints
Type of Joint
Manubriosternal joint: similar in appearance to a cartilaginous symphysis (slightly movable) joint.

Xiphisternal joint: cartilaginous immovable (synchondrosis); usually becomes ossified in old age.

Articulation
Manubriosternal joint: between the manubrium and body of the sternum, adjacent to the second costal cartilage.

Xiphisternal joint: between the body of the sternum and the xiphoid process; this joint marks the inferior extent of the thoracic cavity.

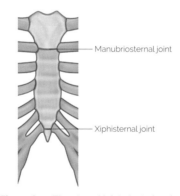

Manubriosternal joint

Xiphisternal joint

Figure 6.15: The sternal joints (anterior view).

Joints of the Shoulder Girdle and Upper Limb

Sternoclavicular Joint
Type of Joint
Functionally, a synovial ball-and-socket joint. Unlike most articular surfaces, however, the articular cartilage is fibrocartilage rather than hyaline cartilage.

Articulation

Between the sternal (medial) end of the clavicle, the clavicular notch of the manubrium, and the costal cartilage of the first rib.

Note: A fibrocartilaginous articular disc separates the joint space into two separate synovial cavities.

Movements

Like other ball and-socket joints, movement occurs in all planes, but anteroposterior movement and rotation is slightly restricted. The joint is involved in the collective movements of the shoulder girdle.

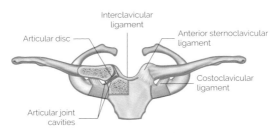

Figure 6.16: The sternoclavicular joint (anterior view). *Note*: The posterior aspect of the joint has a posterior sternoclavicular ligament similar to, but weaker than, the anterior sternoclavicular ligament.

Acromioclavicular Joint

Type of Joint

Synovial plane.

Articulation

Between the lateral end of the clavicle, and the medial border of the acromion of the scapula.

Note: A fibrocartilaginous articular disc partially divides the articular cavity, although it is sometimes absent.

Movements

The joint is involved in the collective movements of the shoulder girdle, enabling the scapula to change its position in relation to the clavicle.

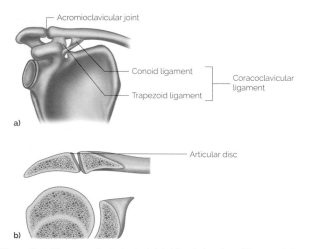

Figure 6.17: The acromioclavicular joint: (a) anterior view; (b) coronal view.

Shoulder (Glenohumeral) Joint

Type of Joint

Synovial ball-and-socket.

Articulation

The head of the humerus articulates with the shallow pear-shaped glenoid cavity (fossa) of the scapula.

This articulation is inherently unstable because the glenoid cavity is only approximately one-third the size of the humeral head, although it is slightly deepened by a rim of fibrocartilage called the *glenoid labrum* or *labrum glenoidale*

(triangular in cross section). The shoulder joint is the most freely moving joint of the body, precisely because stability has been sacrificed to enable maximum range of movement.

Articular Capsule
The articular capsule extends from the margin of the glenoid cavity (including part of the labrum) to the anatomical neck of the humerus. This thin capsule is very loose, thus enabling maximum movement of the joint. When the arm is by the side, the lower part of the capsule hangs in a loose fold, which becomes progressively tauter as the arm is abducted, and increasingly so if the arm continues into elevation. The capsule contributes very little to the stability of the joint. The surrounding muscles, whose attachments are intimately related to the capsule, largely provide joint stability.

Ligaments
Transverse humeral ligament: spans the gap between the humeral tubercles; it holds the long head of the biceps brachii in the intertubercular sulcus as it leaves the joint.

Glenohumeral ligament: three slightly thickened bands of longitudinal fibers on the internal surface of the anterior part of the capsule; may be absent.

Coracohumeral ligament: extends from the coracoid process of the scapula to the upper part of the anatomical neck of the humerus; it greatly reinforces the capsule superiorly and slightly anteriorly.

Coracoacromial ligament: this ligament is completely disconnected from the articular capsule; it forms a shelf above the joint, running between the coracoid process and the acromion process of the scapula.

There are various bursae associated with the shoulder joint. The most important is the subacromial bursa, which separates the coracoacromial ligament from the supraspinatus tendon located above the shoulder joint.

Stabilizing Tendons

Long head of biceps brachii tendon: runs from the superior aspect of the glenoid labrum to enter and travel within the joint cavity, thus proceeding within the articular capsule (hence it is covered with a sheath of synovial membrane). On leaving the cavity, the tendon enters the intertubercular groove of the humerus. Its location secures the head of the humerus tightly against the glenoid cavity, thereby acting as a steadying influence during movements of the shoulder joint.

Rotator cuff tendons: the four rotator cuff tendons (supraspinatus, infraspinatus, teres minor, and subscapularis) encircle the joint and fuse with the articular capsule. Consequently, the rotator cuff muscles or tendons are prone to injury if the joint is vigorously circumducted, as in throwing a ball.

Note: Because overall, the reinforcements of the shoulder joint are weakest inferiorly, the humerus is more prone to dislocate downward.

Movements

Flexion, extension, abduction, adduction, medial and lateral rotation, circumduction, and elevation through flexion and abduction (see Chapter 1).

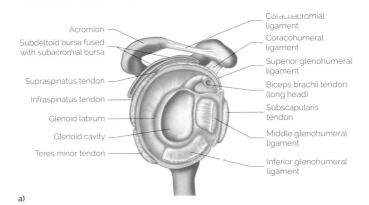

Coracoacromial ligament
Coracohumeral ligament
Superior glenohumeral ligament
Biceps brachii tendon (long head)
Subscapularis tendon
Middle glenohumeral ligament
Inferior glenohumeral ligament

Acromion
Subdeltoid bursa fused with subacromial bursa
Supraspinatus tendon
Infraspinatus tendon
Glenoid labrum
Glenoid cavity
Teres minor tendon

a)

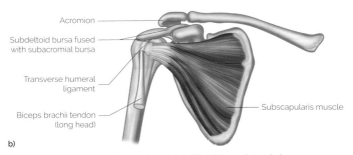

Acromion
Subdeltoid bursa fused with subacromial bursa
Transverse humeral ligament
Biceps brachii tendon (long head)
Subscapularis muscle

b)

Figure 6.18: The shoulder joint: (a) right arm, lateral view; (b) right arm, anterior view (cut).

Elbow Joint

Type of Joint
Synovial hinge (ginglymus).

Articulation
Upper surface of the head of the radius articulates with the capitulum of the humerus. The trochlear notch of the ulna articulates with the trochlea of the humerus (which constitutes the "hinge" mechanism and the main stabilizing factor).

Articular Capsule

The relatively loose articular capsule extends from the coronoid and olecranon fossae of the humerus to the coronoid and olecranon processes of the ulna, and to the anular ligament enclosing the head of the radius. The capsule is thin anteriorly and posteriorly in order to allow flexion and extension, but is strengthened on each side by collateral ligaments.

Ligaments

Ulnar (medial) collateral ligament: three strong bands reinforcing the medial side of the capsule.
Radial (lateral) collateral ligament: a strong triangular ligament reinforcing the lateral side of the capsule.

Stabilizing Tendons

The tendons of the biceps brachii, triceps brachii, and brachialis, as well as many muscles located on the forearm; these tendons cross the elbow joint and provide extra security.

Movements

Flexion and extension only.

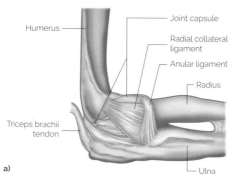

a)

Figure 6.19: (continued)

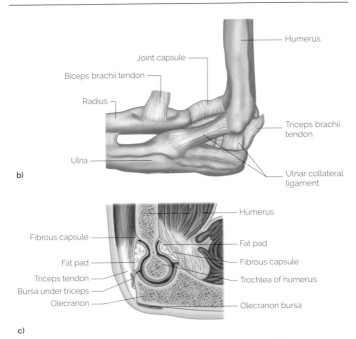

Figure 6.19: The elbow joint: (a) right arm, lateral view; (b) right arm, medial view; (c) right arm, mid-sagittal view.

Proximal Radioulnar Joint

Type of Joint

Synovial pivot.

Articulation

The disc-shaped head of the radius rotates within a ring formed by the radial notch on the ulna and the anular ligament of the radius.

Note: The synovial cavity of this joint is continuous with that of the elbow joint.

Movements
Pronation and supination of the forearm.

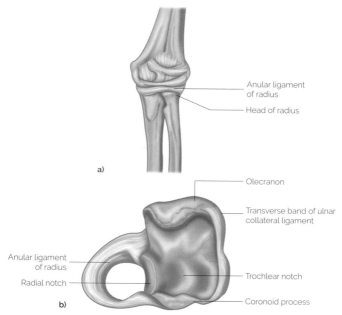

Figure 6.20: The proximal (superior) radioulnar joint: (a) left arm, anterior view; (b) left arm, superior view.

Distal Radioulnar Joint
Type of Joint
Synovial pivot.

Articulation
Between the head of the ulna, and the ulnar notch of the radius.

Note: A fibrocartilaginous articular disc unites the styloid process of the ulna and the medial side of the distal radius.

Movements
Pronation and supination of the forearm.

Intermediate Radioulnar Joint
Type of Joint
Syndesmosis.

Articulation
Connects the interosseous border of the radius with the interosseous border of the ulna, via the interosseous membrane. In addition, a slender fibrous band called the *oblique cord* connects the ulna tuberosity to the proximal end of the shaft of the radius.

Function
Increases the surface of the origin of the deep forearm muscles; helps bind the radius and ulna together; and transmits to the ulna any force passing upward from the hand along the radius.

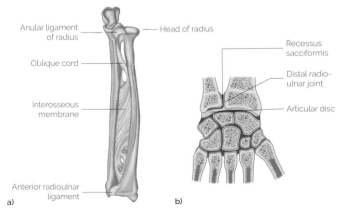

Figure 6.21: The distal and intermediate radioulnar joints: (a) left arm/hand, anterior view; (b) left arm/hand, coronal view.

Radiocarpal Joint (Wrist Joint)
Type of Joint
Synovial ellipsoid.

Articulation
The distal surface of the radius and the articular disc (the same disc as described under "Distal Radioulnar Joint") articulate with the proximal row of carpals, namely the scaphoid, lunate, and triquetral (triquetrum).

Movements
Movements are in combination with the intercarpal joints: flexion, extension, adduction (ulnar deviation), abduction (radial deviation), and circumduction.

Intercarpal Joints
Type of Joint
A series of synovial plane joints.

Articulation
Between the two carpal rows (midcarpal joint), between adjacent bones of the proximal carpal row, and between adjacent bones of the distal carpal row.

Movements
Movements are in combination with the radiocarpal joint: flexion, extension, adduction (ulnar deviation), abduction (radial deviation), and circumduction.

Carpometacarpal Joint of the Thumb
Type of Joint
Synovial saddle joint.

Articulation
Between the trapezium and the base of the first metacarpal bone (the thumb).

Movements
Flexion, extension, abduction, and adduction. At the extreme range of flexion, the first metacarpal medially rotates so that the palmar surface of the thumb becomes opposed to the pads of the fingers. Conversely, slight lateral rotation occurs when the thumb approaches full extension. Combining these movements creates approximate circumduction of the thumb.

Common Carpometacarpal Joint
Type of Joint
Synovial plane.

Articulation
Between the distal row of carpal bones and the bases of the medial four metacarpal bones of the hand.

Movements
Very little movement is possible. However, the articulation at the fifth metacarpal with the hamate is a flattened saddle joint, allowing slight opposition of the little finger across the palm.

Intermetacarpal Joints
Type of Joint
Synovial plane.

Articulation
Between the adjacent sides of the bases of the 2nd to 5th metacarpal bones.

Movements
Limited movement between adjacent metacarpals.

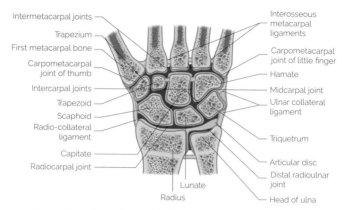

Intermetacarpal joints

Trapezium

First metacarpal bone

Carpometacarpal joint of thumb

Intercarpal joints

Trapezoid

Scaphoid

Radio-collateral ligament

Capitate

Radiocarpal joint

Lunate

Radius

Interosseous metacarpal ligaments

Carpometacarpal joint of little finger

Hamate

Midcarpal joint

Ulnar collateral ligament

Triquetrum

Articular disc

Distal radioulnar joint

Head of ulna

Figure 6.22: The radiocarpal (wrist), intercarpal, carpometacarpal, and intermetacarpal joints (coronal view).

Metacarpophalangeal Joints
Type of Joint
Synovial condyloid.

Articulation
Between the head of a metacarpal and the base of a proximal phalanx.

Note: The capsule is deficient on the dorsal aspect, where it is replaced by an expansion of the long extensor tendon.

Movements
Flexion and extension. Abduction and adduction (possible only in extension, but with very little movement at the thumb). Combined movements may produce circumduction.

Interphalangeal Joints
Type of Joint
Synovial hinge.

Articulation
Between the proximal and middle phalanges (proximal interphalangeal joint, abbreviated PIP), or the middle and distal phalanges (distal interphalangeal joint, abbreviated DIP).

Movements
Flexion and extension.

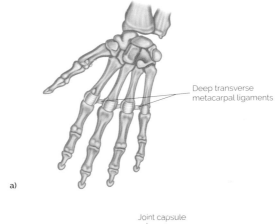

a)

Deep transverse metacarpal ligaments

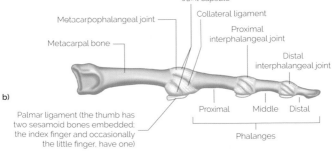

b)

Joint capsule

Collateral ligament

Metacarpophalangeal joint

Proximal interphalangeal joint

Metacarpal bone

Distal interphalangeal joint

Palmar ligament (the thumb has two sesamoid bones embedded; the index finger and occasionally the little finger, have one)

Proximal   Middle   Distal

Phalanges

Figure 6.23: The metacarpophalangeal and interphalangeal joints: (a) anterior view; (b) medial view.

Joints of the Pelvic Girdle and Lower Limb

Lumbosacral and Sacrococcygeal Joints
Type of Joint
Both joints: cartilaginous symphysis (slightly movable).

Articulation
Lumbosacral: between the fifth lumbar vertebra (L5) and the body of the first sacral segment (S1); this joint has the same features as other typical intervertebral joints, with the addition of the iliolumbar ligament.

Sacrococcygeal: between the last sacral and first coccygeal segments; it is reinforced on all sides by the sacrococcygeal ligaments.

Note: Both joints contain a fibrocartilaginous intervertebral disc.

Movements
Lumbosacral joint: contributes to the collective movements of the lumbar vertebral joints.

Sacrococcygeal joint: very little functional movement, and often partially or fully obliterated in old age.

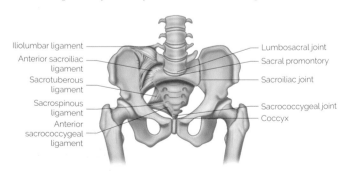

Iliolumbar ligament — Lumbosacral joint
Anterior sacroiliac ligament — Sacral promontory
Sacrotuberous ligament — Sacroiliac joint
Sacrospinous ligament — Sacrococcygeal joint
Anterior sacrococcygeal ligament — Coccyx

Figure 6.24: The lumbosacral, sacroiliac, and sacrococcygeal joints (anterior view).

Sacroiliac Joint

Type of Joint

Synovial joint with pronounced irregular depressions and tubercles on the articular surfaces.

Note: The articular surface of the sacrum is hyaline cartilage, but that of the ilium is usually of the fibrous type.

Articulation

Between the articular surfaces on the sacrum and the iliac bone.

Movements

Very limited movements occur because of the irregular joint surfaces and the strong sacroiliac ligaments.

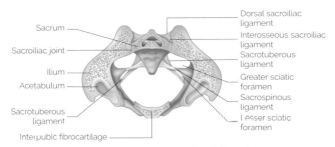

Figure 6.25: Transverse section of the pelvis.

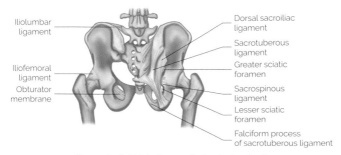

Figure 6.26: Pelvic ligaments (posterior view).

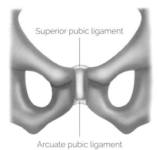

Superior pubic ligament

Arcuate pubic ligament

Figure 6.27: Pubic symphysis (anterior view).

Pubic Symphysis
Type of Joint
Cartilaginous symphysis (slightly movable).

Articulation
Midline joint between the superior rami of the pubic bones.

Note: The joint contains a fibrocartilaginous interpubic disc with a slit-like cavity, which in women can develop into a large cavity.

Movements
No significant movement occurs, other than some separation of the pubic bones in women during pregnancy and childbirth.

Hip Joint
Type of Joint
Synovial ball-and-socket.

Articulation
The spherical head of the femur articulates with the cup-like acetabulum of the coxal (hip) bone. A circular rim of fibrocartilage called the *acetabular labrum* or *labrum acetabulare*, which grasps the femoral head, enhances the depth of the acetabulum. Unlike the articulation of the shoulder joint, the hip articulation fits securely together.

Articular Capsule
The articular capsule extends from the rim of the acetabulum to the neck of the femur. It is very strong and tense in extension, which contrasts with the thin and lax capsule of the shoulder joint.

Ligaments
Iliofemoral ligament: a thick and strong triangular band situated anteriorly.

Pubofemoral ligament: a triangular thickening of the inferior aspect of the capsule.

Ischiofemoral ligament: a spiral ligament situated posteriorly.

These three ligaments are arranged so that when a person stands up (i.e. the hip joint moves from flexion to extension), the head of the femur is "screwed" into the acetabulum and held firmly in position.

Ligament of the head of the femur: also called the *ligamentum teres* or the *capitate ligament*, this flat intracapsular ligament runs from the femoral head to the lower lip of the acetabulum. It contains an artery that is a source of blood for the head of the femur. The ligament is slack during most hip movements and therefore does not contribute to the joint's stability.

Stabilizing Tendons
This joint is inherently stable by virtue of its structure and ligaments. All surrounding muscles and tendons contribute to its stability, but in a very minor capacity compared with the muscles of the shoulder joint.

Movements
Flexion, extension, abduction, adduction, medial and lateral rotation, circumduction (limited, compared with the shoulder joint).

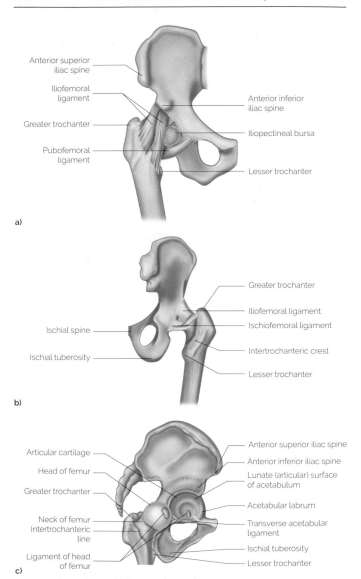

Figure 6.28: The hip joint: (a) right leg, anterior view; (b) right leg, posterior view; (c) right leg, lateral view.

Knee Joint

The knee joint is the largest and most complex joint in the body. Its joint cavity contains three articulations: the lateral and medial articulations of the tibiofemoral joint, and the articulation of the femoropatellar joint.

Type of Joint

Tibiofemoral joint: functionally a modified synovial hinge joint, but structurally a condyloid joint.

Femoropatellar joint: synovial plane joint.

Articulation

Tibiofemoral joint: the condyles of the femur articulate with the condyles of the tibia, but with two C-shaped menisci or semilunar cartilages between the opposing articular surfaces.

Femoropatellar joint: the posterior surface of the patella articulates with the patellar surface at the lower end of the femur.

Articular Capsule

The knee is the only joint where the capsule only partially encloses the joint cavity. Instead, the true capsular fibers are integrated within a ligamentous sheath composed of muscle tendons or expansions from them, which collectively encapsulate the joint. True capsular fibers are located only at the sides and posterior of the joint.

Extracapsular (extra-articular) Ligaments

Tibial (medial) collateral ligament: a broad, flat band running from the medial epicondyle of the femur, downward and forward to the medial condyle of the tibial shaft; some fibers are fused to the medial meniscus.

Fibular (lateral) collateral ligament: a round, cord-like ligament, fully detached from the thin lateral part of the capsule; it extends from the lateral epicondyle of the femur, downward and backward to the head of the fibula.

Oblique popliteal ligament: an expansion of the semimembranosus tendon, which passes upward and laterally over the posterior of the joint.

Arcuate popliteal ligament: extends from the head of the fibula upward and medially, spreading into the back of the capsule and to the lateral condyle of the femur, thus reinforcing the back of the joint.

Intracapsular (intra-articular) Ligaments and Menisci
Anterior cruciate ligament: extends obliquely upward, laterally, and backward from the anterior intercondylar area of the tibia to the medial surface of the lateral femoral condyle; it prevents posterior displacement of the femur on the tibia, and also helps check hyperextension of the knee.

Posterior cruciate ligament: passes upward, medially, and forward from the posterior intercondylar area of the tibia to the lateral side of the medial femoral condyle. The ligament therefore lies on the medial side of the weaker anterior cruciate; it prevents anterior displacement of the femur on the tibia.

The cruciate ligaments lie within the joint capsule, but outside the joint cavity. The synovial membrane covers most of their surfaces.

Menisci: two crescent-shaped fibrous wedges between the femoral and tibial condyles. They help compensate for the incongruence of the articular surfaces, and also help absorb shock transmitted to the knee joint. The menisci are attached only at their outer margins and are prone to tearing. The medial meniscus is also attached to the tibial collateral ligament, and is therefore much more firmly anchored than the lateral meniscus, which does not attach to the fibular collateral ligament.

Medial and lateral coronary ligaments: capsular fibers that attach the menisci to the tibial condyles.

Transverse ligament of the knee: a fibrous band that joins the anterior parts of the menisci.

Stabilizing Tendons

Patellar ligament (ligamentum patellae): a strong ligament that is actually the distal part of the quadriceps tendon. The ligament runs from the patella (which is embedded within the tendon as a sesamoid bone—see p. 137) to the tibial tuberosity.

Other thinner bands, called the *medial patellar retinaculum* and the *lateral patellar retinaculum*, pass down the sides of the patella to attach to the front of each tibial condyle, effectively substituting for the capsule anteriorly.

Semimembranosus tendon: helps reinforce the posterior of the knee joint.

The muscles surrounding the knee joint are particularly crucial as joint stabilizers.

Movements

Flexion, extension. Some rotation can occur when the knee is flexed. Moreover, as a result of the tightening of various ligaments (especially the cruciates) and tendons, slight medial rotation of the femur occurs upon the fixed tibia as the knee straightens into full extension. (When both the femur and the tibia are not fixed, as in kicking actions, the tibia rotates laterally at the end of extension and medially at the beginning of flexion).

Note: The popliteus muscle "unlocks" the extended knee joint prior to flexion, enabling flexion to occur.

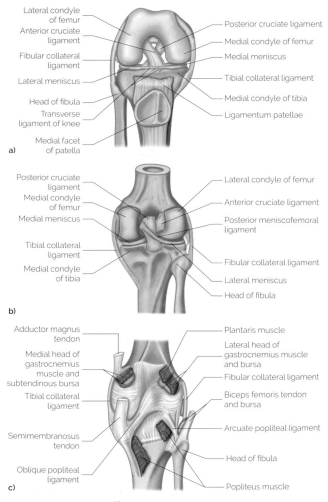

a)
Lateral condyle of femur
Anterior cruciate ligament
Fibular collateral ligament
Lateral meniscus
Head of fibula
Transverse ligament of knee
Medial facet of patella
Posterior cruciate ligament
Medial condyle of femur
Medial meniscus
Tibial collateral ligament
Medial condyle of tibia
Ligamentum patellae

b)
Posterior cruciate ligament
Medial condyle of femur
Medial meniscus
Tibial collateral ligament
Medial condyle of tibia
Lateral condyle of femur
Anterior cruciate ligament
Posterior meniscofemoral ligament
Fibular collateral ligament
Lateral meniscus
Head of fibula

c)
Adductor magnus tendon
Medial head of gastrocnemius muscle and subtendinous bursa
Tibial collateral ligament
Semimembranosus tendon
Oblique popliteal ligament
Plantaris muscle
Lateral head of gastrocnemius muscle and bursa
Fibular collateral ligament
Biceps femoris tendon and bursa
Arcuate popliteal ligament
Head of fibula
Popliteus muscle

Figure 6.29: (continued)

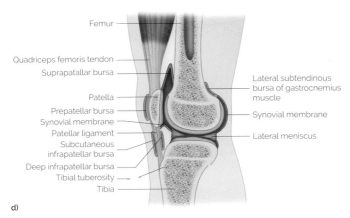

Femur

Quadriceps femoris tendon

Suprapatallar bursa

Patella

Prepatellar bursa

Synovial membrane

Patellar ligament

Subcutaneous infrapatellar bursa

Deep infrapatellar bursa

Tibial tuberosity

Tibia

Lateral subtendinous bursa of gastrocnemius muscle

Synovial membrane

Lateral meniscus

d)

Figure 6.29: The knee joint. (a) right leg, anterior view; (b) right leg, posterior view; c) right leg, posterior view; d) right leg, mid-sagittal view.

Proximal Tibiofibular Joint
Type of Joint
Synovial plane.

Articulation
Between a facet on the head of the fibula and a similar facet on the lateral condyle of the tibia.

Movements
Movement is slight and passively occurs along with movements of the ankle joint.

Distal Tibiofibular Joint
Type of Joint
Syndesmosis.

Articulation
Between the rough, triangular, opposed surfaces at the distal ends of the tibia and fibula.

Movements
Movement is slight and passively occurs along with movements of the ankle joint.

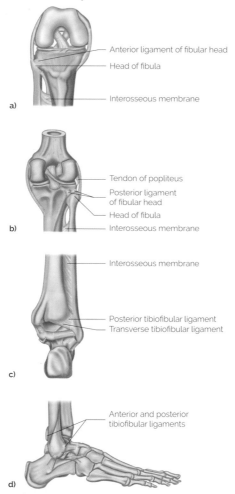

a) Anterior ligament of fibular head

Head of fibula

Interosseous membrane

b) Tendon of popliteus

Posterior ligament of fibular head

Head of fibula

Interosseous membrane

c) Interosseous membrane

Posterior tibiofibular ligament
Transverse tibiofibular ligament

d) Anterior and posterior tibiofibular ligaments

Figure 6.30: The tibiofibular joints: (a) proximal tibiofibular joint, right leg, anterior view; (b) proximal tibiofibular joint, right leg, posterior view; (c) distal tibiofibular joint, right leg, posterior view; (d) distal tibiofibular joint, right leg, lateral view.

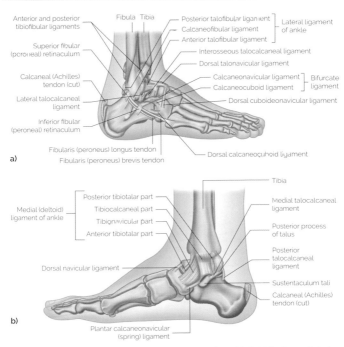

Figure 6.31: The ankle joint: (a) right foot, lateral view; (b) right foot, medial view.

Ankle Joint

Type of Joint
Synovial hinge.

Articulation
Between the distal tibia, the medial malleolus of the tibia, the lateral malleolus of the fibula, and the talus. Therefore, the lower ends of the tibia and fibula provide a socket for the talus.

Movements
Dorsiflexion and plantar flexion.

The Arches of the Foot

The bones of the foot arrange themselves into three distinct arches: the lateral longitudinal, the medial longitudinal, and the transverse. Their shape allows them to act like a spring, bearing the weight of the body, as well as absorbing shock produced by movement. The flexibility of these arches facilitates functions such as walking and running.

The lateral longitudinal arch is made up of four bones: the calcaneus, cuboid, and fourth and fifth metatarsals. It is supported by fibularis longus, flexor digitorum longus, flexor hallucis, as well as the intrinsic foot muscles; plantar ligaments; the shape of the bones of the arch; plantar aponeurosis.

The medial longitudinal arch is made up of the talus, calcaneus, navicular, three cuneiforms, and three metatarsals. This arch bears most of the body's weight. It is supported by tibialis anterior and posterior, fibularis longus, flexor

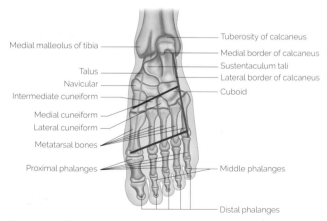

Figure 6.32: The bones of the foot arrange themselves into three distinct arches: the lateral longitudinal, the medial longitudinal, and the transverse (proximal and distal).

digitorum longus, flexor hallucis, as well as the intrinsic foot muscles; plantar ligaments, medial ligaments of the ankle; the shape of the bones of the arch; plantar aponeurosis.

The transverse arch is located on the coronal plane, and comprises the three cuneiforms, the cuboid, and the metatarsal bases. It is supported by fibularis longus and tibialis posterior; plantar ligaments and deep transverse metatarsal ligaments; the wedged shape of the bones of the arch; plantar aponeurosis.

Intertarsal Joints
Type of Joint
Complex set of synovial plane joints.

Articulation
Subtalar joint: between the inferior surface of the talus and the superior surface of the calcaneus.

Talocalcaneonavicular joint: between the talus, calcaneus, and navicular.

Calcaneocuboid joint: between the calcaneus and cuboid.

Transverse tarsal joint: a term used to describe the transverse plane extending across the full width of the tarsus, comprising the talocalcaneonavicular joint and the calcaneocuboid joint.

Cuneonavicular joint: between the cuneiform and the navicular.

Intercuneiform joints: between the three cuneiform bones.

Cuneocuboid joint: between the lateral cuneiform bone and the cuboid bone.

Movements
Tarsus: inversion and eversion of the foot.

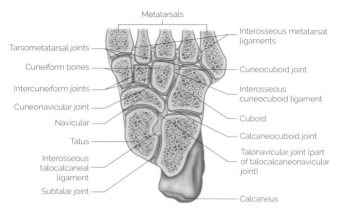

Figure 6.33: The intertarsal joints (horizontal section of the right foot).

Tarsometatarsal and Intermetatarsal Joints
Type of Joint
Synovial plane.

Articulation
Tarsometatarsal joints: between the distal (anterior) row of tarsal bones (the cuboid and three cuneiforms) and the bases of the metatarsal bones.

Intermetatarsal joints: between the facets on adjacent sides of the bases of all lateral metatarsal bones.

Movements
Small gliding movements of the metatarsals, limited by ligaments and the interlocking of the bones, contribute slightly to inversion and eversion of the foot.

Metatarsophalangeal Joints
Type of Joint
Synovial condyloid.

Articulation
Between the head of a metatarsal and the base of a proximal phalanx.

Note: The capsule is deficient on the dorsal aspect, where it is replaced by an expansion of the extensor tendon.

Movements
Flexion and extension. Abduction and adduction. Combined movements may produce passive circumduction.

Note: In flexion, the toes are drawn together; in extension they tend to spread apart and incline slightly laterally. Movements are less extensive than at the corresponding joints of the hand.

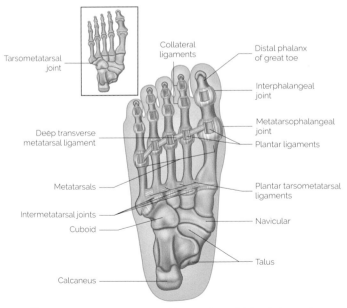

Figure 6.34: The tarsometatarsal, intermetatarsal, interphalangeal, and metatarsophalangeal joints (plantar view).

Interphalangeal Joints
Type of Joint
Synovial hinge.

Articulation
Between the proximal and middle phalanges (proximal interphalangeal joint, abbreviated PIP), or the middle and distal phalanges (distal interphalangeal joint, abbreviated DIP).

Movements
Flexion and extension.

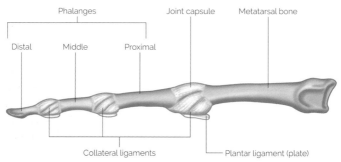

Figure 6.35: The metatarsophalangeal and interphalangeal joints (lateral view).

CHAPTER 7

The Musculoskeletal System

Structure and Function of Skeletal Muscle

Skeletal (somatic or voluntary) muscles make up approximately 40% of the total human body weight. Their primary function is to produce movement through the ability to contract and relax in a coordinated manner. They are attached to bone either directly or more often via tendons. The location where a muscle attaches to a relatively stationary point on a bone, either directly or via a tendon, is called the *origin*. When the muscle contracts, it transmits tension to the bones across one or more joints, and movement occurs. The end of the muscle that attaches to the bone that moves is called the *insertion*.

Overview of Skeletal Muscle Structure

The functional unit of skeletal muscle is known as a *muscle fiber*, which is an elongated, cylindrical cell with multiple nuclei, ranging from 10 to 100 microns in width, and a few millimeters to 30+ centimeters in length. The cytoplasm of the fiber is called the *sarcoplasm*, which is encapsulated inside a cell membrane called the *sarcolemma*. A delicate membrane known as the *endomysium* surrounds each individual fiber.

The muscle fibers are grouped together in bundles covered by a collagenic sheath known as the *perimysium*. These bundles are themselves grouped together, and the whole muscle is encased in a sheath called the *epimysium*. These muscle membranes lie throughout the entire length of the muscle, from the tendon of origin to the tendon of the insertion. This entire structure is sometimes referred to as the *musculotendinous unit*.

In defining the structure of muscle tissue in more detail, from microscopic to gross anatomy, we therefore have the following components: myofibrils, endomysium, fasciculi, perimysium, epimysium, deep fascia, and superficial fascia.

Myofibrils

Through an electron microscope, one can distinguish the contractile elements of a muscle fiber, known as *myofibrils*, running the entire length of the fiber. Each myofibril reveals alternate light and dark banding, producing the characteristic cross-striation of the muscle fiber; these bands are called *myofilaments*. The light bands are referred to as *isotropic (I) bands*, and consist of thin myofilaments made of the protein actin. The dark ones are called *anisotropic (A) bands*, consisting of thicker myofilaments made of the protein myosin. A third connecting filament is made of the sticky protein titin, also known as connectin, which is the third most abundant protein in human tissue.

The myosin filaments have paddle-like extensions that emanate from them, rather like the oars of a boat. These extensions latch onto the actin filaments, forming what are described as *cross-bridges* between the two types of filaments. These cross-bridges, using the muscle energy source known as *adenosine triphosphate (ATP)*, pull the actin strands closer together.* Thus, the light and

---

*Huxley's Sliding Filament Theory**. The generally accepted hypothesis to explain muscle function is partly described by the sliding filament theory proposed by Huxley and Hanson in 1954. Muscle fibers receive a nerve impulse that causes the release of

dark sets of filaments increasingly overlap, like interlocking fingers, resulting in muscle contraction. A set of actin-myosin filaments is called a *sarcomere*.

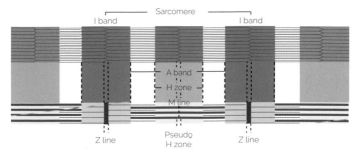

Figure 7.1: The myofilaments in a sarcomere. A sarcomere is bounded at both ends by the Z line.

- The lighter zone is known as the *I band*, and the darker zone the *A band*.
- The *Z line* is a thin dark line at the midpoint of the I band.
- A *sarcomere* is defined as the section of myofibril between consecutive Z lines.
- The center of the A band contains the *H zone*.
- The **M line** bisects the H zone, and delineates the center of the sarcomere.

If an outside force causes a muscle to stretch beyond its resting level of tonus (the slight, continuous contraction of a muscle, aiding the maintenance of posture), the interlinking effect of the actin and myosin filaments that occurs during contraction

---

calcium ions stored in the muscle. In the presence of the ATP, the calcium ions bind with the actin and myosin filaments to form an electrostatic (magnetic) bond. This bond causes the fibers to shorten, resulting in their contraction or an increase in tonus (muscle tone). When the nerve impulse ceases, the muscle fibers relax. Because of their elastic nature, the filaments recoil to their non-contracted lengths, i.e. their resting level of tonus.

is reversed. Initially, the actin and myosin filaments accommodate the stretch, but as the stretch continues, the titin filaments increasingly "pay out" to absorb the displacement. Thus, it is the titin filament that determines the muscle fiber's extensibility and resistance to stretch. Research indicates that a muscle fiber (sarcomere) can be elongated to 150% of its normal length at rest.

Endomysium
A delicate connective tissue called *endomysium* lies outside the sarcolemma of each muscle fiber, separating each fiber from its neighbors, but also connecting them together.

Fasciculi
Muscle fibers are arranged in parallel bundles called *fasciculi*.

Perimysium
Each fasciculus is bound by a denser collagenic sheath called the *perimysium*.

Epimysium
The entire muscle, which is therefore an assembly of fasciculi, is wrapped in a fibrous sheath called the *epimysium*; this arrangement facilitates force transmission.

Deep Fascia
A coarser sheet of fibrous connective tissue lies outside the epimysium, binding individual muscles into functional groups. This deep fascia extends to wrap around other adjacent structures.

Superficial Fascia
While its anatomy and topography differs from region to region, providing specialization, the superficial fascia is primarily a fatty layer that contains oblique septa and

connects the skin to deep fascia. Contractile fibers have been reported in the superficial fascia, particularly in the neck.

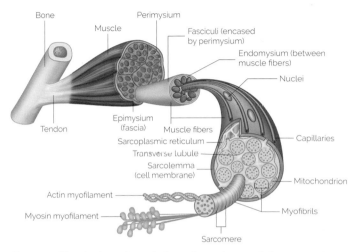

Figure 7.2: The structure of muscle tissue from microscopic to gross anatomy.

Muscle Attachment

The way a muscle attaches to bone or other tissues is through either a direct or an indirect attachment. A *direct or fleshy attachment* is where the perimysium and epimysium of the muscle unite and fuse with the periosteum of bone, perichondrium of cartilage, a joint capsule, or the connective tissue underlying the skin, as in some muscles of facial expression. An *indirect attachment* is where the connective tissue components of a muscle fuse together into bundles of collagen fibers to form an intervening tendon. Indirect attachments are much more common. The different types of indirect attachment are: tendons and aponeuroses, intermuscular septa, and sesamoid bones.

Tendons and Aponeuroses

When the connective tissue components of a muscle combine and extend beyond the end of the muscle as round cords or flat bands, the tendinous attachment is called a *tendon*; if they extend as a thin, flat, and broad sheet-like material, the attachment is called an *aponeurosis*. The tendon or aponeurosis secures the muscle to bone or cartilage, to the fascia of other muscles, or to a seam of fibrous tissue called a *raphé*. Flat patches of tendon may form on the body of a muscle where it is exposed to friction. This may occur, for example, on the deep surface of the trapezius where it rubs against the spine of the scapula.

Intermuscular Septa

In some cases, flat sheets of dense connective tissue known as *intermuscular septa* penetrate between muscles, providing another structure to which muscle fibers may attach.

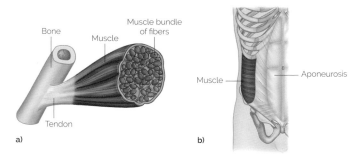

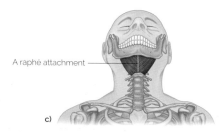

Figure 7.3: (a) Tendon attachment; (b) attachment by aponeurosis; (c) mylohyoid raphé.

Sesamoid Bones

If a tendon is subject to friction, it may, though not in all cases, develop a sesamoid bone within its substance. The largest sesamoid bone in the body is the patella or kneecap. However, sesamoid bones may also appear in tendons not subject to friction.

Multiple Attachments

Many muscles have only two attachments, one at each end. More complex muscles, on the other hand, are often attached to several different structures at their origins and/or their insertions. If these attachments are separated, so that there are two or more tendons and/or aponeuroses inserting into different places, the muscle is said to have two or more heads. For example, the biceps brachii has two heads at its origin: one from the coracoid process of the scapula, and the other from the supraglenoid tubercle. The triceps brachii has three heads and the quadriceps femoris has four.

Red and White Muscle Fibers

Three types of skeletal muscle fibers have been distinguished: (1) red slow-twitch fibers; (2) white fast-twitch fibers; and (3) intermediate fast-twitch fibers. There is always a mixture of these types of muscle fibers in any given muscle, giving them a range of fatigue resistance and contractile speeds.

1. *Red slow-twitch fibers*: These fibers are thin cells that contract slowly. The red color is due to their content of myoglobin, a substance similar to hemoglobin, which stores oxygen and increases the rate of oxygen diffusion within the muscle fibers. As long as the oxygen supply is plentiful, red fibers can contract for sustained periods, and are thus very resistant to fatigue. Successful marathon runners tend to have a high percentage of these red fibers.
2. *White fast-twitch fibers*: These fibers are large cells that contract rapidly. They are pale because of their lower

content of myoglobin. White fibers fatigue quickly, because they rely on short-lived glycogen reserves in the fiber to contract. However, they are capable of generating much more powerful contractions than red fibers, enabling them to perform rapid, powerful movements for short periods. Successful sprinters have a higher proportion of these white fibers.

3. *Intermediate Fast-twitch fibers*: These red or pink fibers are a compromise in size and activity between red and white fibers.

Blood Supply

In general, each muscle receives an arterial supply to bring nutrients via the blood into the muscle, and contains several veins to take away metabolic by-products released by the muscle into the blood. These blood vessels usually enter through the central part of the muscle, but may also enter toward one end. Thereafter, they branch into a capillary plexus, which spreads throughout the intermuscular septa, to eventually penetrate the endomysium around each muscle fiber. During exercise, the capillaries dilate, increasing the amount of blood flow in the muscle by up to 800 times. However, a muscle tendon, being composed of a relatively inactive tissue, has a much less extensive blood supply.

Nerve Supply

The nerve supply to a muscle usually enters at the same place as the blood supply in a neurovascular bundle, and branches through the connective tissue septa into the endomysium in a similar way. Each skeletal muscle fiber is supplied by a single nerve ending. This is in contrast to some other muscle tissues, which are able to contract without any nerve stimulation.

The nerve entering the muscle usually contains roughly equal proportions of sensory and motor nerve fibers, although some muscles may receive separate sensory branches. As the nerve fiber approaches the muscle fiber, it divides into a number of terminal branches, collectively known as a *motor end plate*.

Motor Unit of a Skeletal Muscle

A motor unit consists of a single motor nerve cell and the muscle fibers that it stimulates. Motor units vary in size, ranging from cylinders of muscle 5–7 mm in diameter in the upper limb, to 7–10 mm in diameter in the lower limb. The average number of muscle fibers within a unit is 150, but this number can range from less than ten to several hundred. Where fine gradations of movement are required, as in the muscles of the eyeball or fingers, the number of muscle fibers supplied by a single nerve cell is small. On the other hand, where mass movements are required, as in the muscles of the lower limb, each nerve cell may supply a motor unit of several hundred fibers.

The muscle fibers in a single motor unit are spread throughout the muscle, rather than being clustered together. This means that stimulation of a single motor unit will cause the entire muscle to exhibit a weak contraction.

Skeletal muscles work on an "all or nothing principle": in other words, groups of muscle cells, or fasciculi, can either contract or not at all. Depending on the strength of contraction required, a certain number of muscle cells will fully contract, while others will not. When a greater muscular effort is needed, most of the motor units may be stimulated at the same time. However, under normal conditions, the motor units tend to work in relays, so that during prolonged contractions some are inhibited while others are contracting— this is known as *gradual increments of contraction (GIC)*.

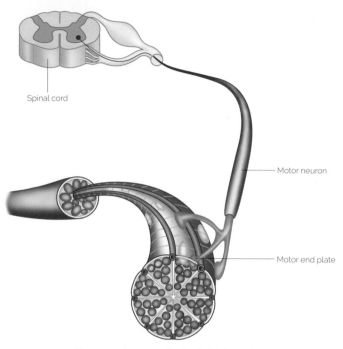

Spinal cord

Motor neuron

Motor end plate

Figure 7.4: A motor unit of a skeletal muscle.

Muscle Reflexes

Within skeletal muscles there are two specialized types
of nerve receptor that can sense tension (length or stretch):
muscle spindles and Golgi tendon organs (GTOs). *Muscle
spindles* are cigar-like in shape and consist of tiny modified
muscle fibers called *intrafusal fibers*, and nerve endings,
encased together within a connective tissue sheath; they
lie between and parallel to the main muscle fibers. *GTOs*
are located mostly at the junctions of muscles and their
tendons or aponeuroses.

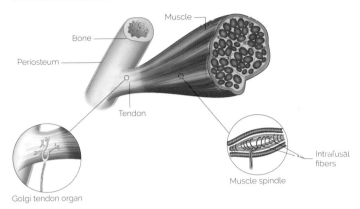

Figure 7.5: Anatomy of the muscle spindle and Golgi tendon organ.

Stretch Reflex

The *stretch reflex* helps control posture by maintaining muscle tone. It also helps prevent injury, by enabling a muscle to respond to a sudden or unexpected increase in length. This is how it works:

1. When a muscle is lengthened, the muscle spindles are excited, causing each spindle to send a nerve impulse communicating the speed of lengthening to the spinal cord.
2. On receiving this impulse, the spinal cord immediately sends a proportionate impulse back to the stretched muscle fibers, causing them to contract, in order to decelerate the movement. This circular process is known as the *reflex arc*.
3. An impulse is simultaneously sent from the spinal cord to the antagonist of the contracting muscle (i.e. the muscle opposing the contraction), inhibiting the action of the antagonist so that it cannot resist the contraction of the stretched muscle. This process is known as *reciprocal inhibition*.
4. Concurrent with this spinal reflex, nerve impulses are also sent up the spinal cord to the brain to relay information about muscle length and the speed of muscle contraction. A reflex in the brain feeds nerve impulses back to the muscles in order

to ensure that appropriate muscle tone is maintained to meet the requirements of posture and movement.

5. Meanwhile, the stretch sensitivity of the minute intrafusal muscle fibers within the muscle spindle are evened out and regulated by gamma efferent nerve fibers*, arising from motor neurons within the spinal cord. Thus, a gamma motor neuron reflex arc ensures the evenness of muscle contraction, which would otherwise be jerky if muscle tone relied solely on the stretch reflex.

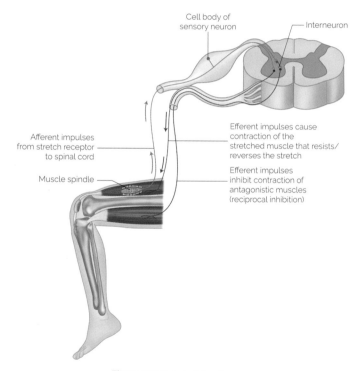

Cell body of sensory neuron

Interneuron

Afferent impulses from stretch receptor to spinal cord

Efferent impulses cause contraction of the stretched muscle that resists/reverses the stretch

Muscle spindle

Efferent impulses inhibit contraction of antagonistic muscles (reciprocal inhibition)

Figure 7.6: The stretch reflex arc.

---

*The function of these nerve fibers is to regulate the sensitivity of the spindle and the total tension in the muscle.

The classic clinical example of the stretch reflex used in clinical practice is the *knee jerk*, or *patellar reflex*, whereby the patellar tendon is lightly struck with a small rubber hammer. This results in the following sequence of events:

1. The sudden stretch of the patellar tendon causes the quadriceps to be stretched, i.e. the sharp tap on the patellar tendon causes a sudden stretch of the tendon.
2. This rapid stretch is registered by the muscle spindles within the quadriceps, causing the quadriceps to contract. This causes a small kick as the knee straightens suddenly, and takes the tension off the muscle spindles.
3. Simultaneously, nerve impulses to the hamstrings, which are the antagonists of the quadriceps, result in functional inhibition of their action.

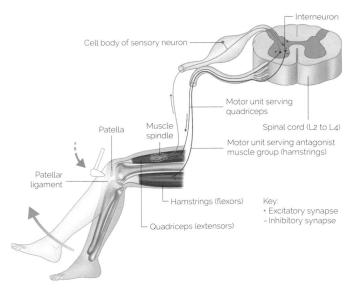

Figure 7.7: The patellar reflex.

Another well-known example of the stretch reflex in action occurs when a person falls asleep in the seated position: their head will relax forward, then jerk back up, because the stretched muscle spindles in the back of the neck have activated a reflex arc.

The stretch reflex also works constantly to maintain the tonus of our postural muscles; in other words, it enables us to remain standing without conscious effort and without collapsing forward. The sequence of events preventing this forward collapse occurs in a fraction of a second, as follows:

1. In standing, we naturally begin to sway forward.
2. This pulls our calf muscles into a lengthened position, activating the stretch reflex.
3. The calf muscles consequently contract to pull us back to the upright position.

Deep Tendon Reflex (Autogenic Inhibition)

In contrast to the stretch reflex, which involves the response of the muscle spindles to elongation of muscle fibers, the *deep tendon reflex* involves the reaction of GTOs to muscle contraction or an undue rise in tension. Accordingly, the deep tendon reflex creates the opposite effect to that of the stretch reflex. This is how it works:

1. When a muscle contracts, it pulls on the tendons which are situated at both ends.
2. The tension in the tendon causes the GTOs to transmit impulses to the spinal cord. Some impulses continue to the cerebellum.
3. As these impulses reach the spinal cord, they inhibit the motor nerves supplying the contracting muscle, thus reducing tonus.

4. Simultaneously, the motor nerves supplying the antagonist muscle are activated, causing it to contract. This process is called *reciprocal activation*.
5. Meanwhile, the information reaching the cerebellum is processed and fed back to help readjust muscle tension.

The deep tendon reflex has a protective function: it prevents the muscle from contracting so hard that it would pull its attachment off the bone. It is therefore especially important during activities which involve rapid switching between flexion and extension, such as running.

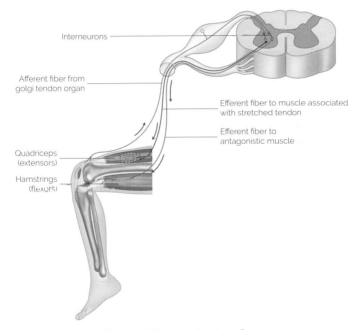

Interneurons

Afferent fiber from golgi tendon organ

Efferent fiber to muscle associated with stretched tendon

Efferent fiber to antagonistic muscle

Quadriceps (extensors)

Hamstrings (flexors)

Figure 7.8: The deep tendon reflex.

Note, however, that in normal day-to-day movement, tension in the muscles is not sufficient to activate the GTOs and cause a deep tendon reflex. By contrast, the threshold of the muscle spindle stretch reflex is set much lower, because it must constantly maintain sufficient tonus in the postural muscles to keep the body upright.

Isometric and Isotonic Contractions

A muscle will contract upon stimulation in an attempt to bring its attachments closer together, but this does not necessarily result in a shortening of the muscle. If the contraction of a muscle results in no movement, such a contraction is called *isometric*; if movement of some sort results, the contraction is called *isotonic*.

Isometric Contraction

An *isometric* contraction occurs when there is increased tension in a muscle, but its length remains unchanged. In other words, although the muscle tenses, the joint over which the muscle passes does not move. One example of this is holding a heavy object in the hand with the elbow held stationary and bent at 90 degrees. Trying to lift something that proves to be too heavy to move is another example. Note also that some of the postural muscles are largely working isometrically by automatic reflex. For example, in the upright position, the body has a natural tendency to fall forward at the ankle; this is prevented by isometric contraction of the calf muscles. Likewise, the center of gravity of the skull would make the head tilt forward if the muscles at the back of the neck did not contract isometrically to keep the head centralized.

Isotonic Contraction

Isotonic contractions of muscle enable us to move about. Such contractions are of two types: concentric and eccentric.

In *concentric* contractions, the muscle attachments move closer together, causing movement at the joint. In the example of holding an object in the hand, if the biceps muscle contracts concentrically, the elbow joint will flex and the hand will move toward the shoulder. Similarly, if we look up at the ceiling, the muscles at the back of the neck must contract concentrically to tilt the head back and extend the neck.

Eccentric contraction means that the muscle fibers "pay out" in a controlled manner to slow down movements in a case where gravity, if unchecked, would otherwise cause them to occur too rapidly, as, for example, when lowering an object held in the hand down to your side. Another everyday example is simply sitting down onto a chair. Therefore, the difference between concentric and eccentric contractions is that in the former, the muscle shortens, while in the latter, it actually lengthens.

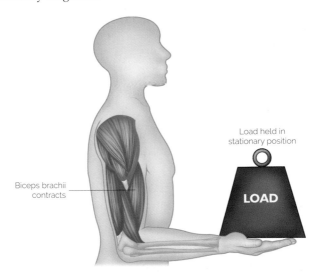

Load held in
stationary position

Biceps brachii
contracts

LOAD

Figure 7.9: Isometric contraction.

Figure 7.10: Abdominal muscles contract concentrically to raise the body.

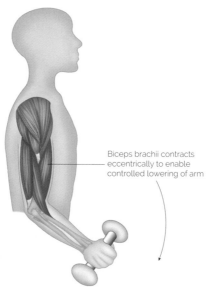

Biceps brachii contracts
eccentrically to enable
controlled lowering of arm

Figure 7.11: Eccentric isotonic contraction.

Muscle Shape (Arrangement of Fascicles)

Muscles come in a variety of shapes according to the arrangement of their fascicles. The reason for this variation is to provide optimum mechanical efficiency for a muscle in relation to its position and action. The most common arrangement of fascicles yields muscle shapes which can be described as parallel, pennate, convergent, and circular, with each of these shapes having further sub-categories. The different shapes are illustrated in Figure 7.12.

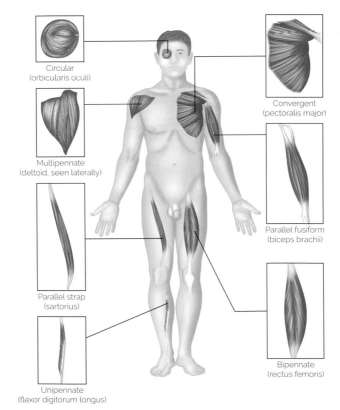

Circular
(orbicularis oculi)

Multipennate
(deltoid, seen laterally)

Parallel strap
(sartorius)

Unipennate
(flexor digitorum longus)

Convergent
(pectoralis major)

Parallel fusiform
(biceps brachii)

Bipennate
(rectus femoris)

Figure 7.12: Muscle shapes.

Parallel

In this arrangement the fascicles are arranged parallel to the long axis of the muscle. If the fascicles extend throughout the length of the muscle, it is known as a *strap muscle*, as, for example, the sartorius. If the muscle also has an expanded belly and tendons at both ends, it is called a *fusiform muscle*, as, for example, the biceps brachii. A variation of this type of muscle has a fleshy belly at either end, with a tendon in the middle; as in the digastric muscle.

Pennate

Pennate muscles are so named because their short fasciculi are attached obliquely to the tendon, like the structure of a feather (from Latin *penna* = "feather"). If the tendon develops on one side of the muscle, it is referred to as *unipennate*, as in, for example, the flexor digitorum longus in the leg. If the tendon is in the middle and the fibers are attached obliquely from both sides, it is known as *bipennate*, a good example being the rectus femoris. If there are numerous tendinous intrusions into the muscle, with fibers attaching obliquely from several directions (thus resembling many feathers side by side), the muscle is referred to as *multipennate*; the best example is the deltoid muscle.

Convergent

Muscles that have a broad origin with fascicles converging toward a single tendon, giving the muscle a triangular shape, are called *convergent muscles*. The best example is the pectoralis major.

Circular

When the fascicles of a muscle are arranged in concentric rings, the muscle is referred to as *circular*. All the sphincter skeletal muscles in the body are of this type; they surround openings, which they close by contracting. An example is the orbicularis oculi.

Range of Motion Versus Power

When a muscle contracts, it can shorten by up to 70% of its original length; hence, the longer the fibers, the greater the range of motion. On the other hand, the strength of a muscle depends on the total number of muscle fibers it contains, rather than the length of the fibers. Therefore:

1. Muscles with long parallel fibers produce the greatest range of motion, but are not usually very powerful.
2. Muscles with a pennate pattern (especially multipennate) pack in the most fibers. Such muscles shorten less than long parallel muscles, but tend to be much more powerful.

Functional Characteristics of a Skeletal Muscle

All that has been said about muscles so far in this book enables us to formulate a list of functional characteristics pertaining to skeletal muscle.

Excitability

Excitability is the ability to receive and to respond to a stimulus. In the case of a muscle, when a nerve impulse from the brain reaches the muscle, a chemical known as *acetylcholine* is released. This chemical produces a change in the electrical balance in the muscle fiber and, as a result, generates an electrical current known as an *action potential*. The action potential conducts the electrical current from one end of the muscle cell to the other and results in a contraction of the muscle cell, or muscle fiber (remember that one muscle cell = one muscle fiber).

Contractility

Contractility is the ability of a muscle to shorten forcibly when stimulated. The muscles themselves can only contract; they do not usually lengthen, except via some external means (i.e. manually), beyond their normal resting length (see "Tonus" below). In other words, muscles can only pull their ends together (contract); they cannot push them apart.

Extensibility

Extensibility is the ability of a muscle to be extended, or returned to its resting length (which is a semi-contracted state) or slightly beyond. For example, if we bend forward at the hips from standing, the muscles of the back, such as the erector spinae, lengthen eccentrically (see p. 147) to lower the trunk, paying out slightly beyond their normal resting length, and are thus effectively "elongated."

Elasticity

Elasticity describes the ability of a muscle fiber to recoil after being lengthened, and therefore resume its resting length when relaxed. In a whole muscle, the elastic effect is supplemented by the important elastic properties of the connective tissue sheaths (endomysium and epimysium). Tendons also contribute some elastic properties. An example of this elastic recoil effect can be experienced when coming back up from a forward bend at the hips as described above. Initially there is no muscle contraction; instead, the upward movement is initiated purely by elastic recoil of the back muscles, after which the contraction of the back muscles completes the movement.

Tonus

Tonus, or *muscle tone*, is the term used to describe the slightly contracted state to which muscles return during the resting state. Muscle tonus does not produce active movements, but it keeps the muscles firm, healthy, and ready to respond to stimulation. It is the tonus of skeletal muscles that also helps stabilize and maintain posture. *Hypertonic muscles* are those muscles whose "normal" resting state is over-contracted.

General Functions of Skeletal Muscles

- **Enable movement:** skeletal muscles are responsible for all locomotion and manipulation, and they enable you to respond quickly.
- **Maintain posture:** skeletal muscles support an upright posture against the pull of gravity.
- **Stabilize joints:** skeletal muscles and their tendons stabilize joints.
- **Generate heat:** skeletal muscles (in common with smooth and cardiac muscles) generate heat, which is important in maintaining a normal body temperature.

Musculoskeletal Mechanics

Origins and Insertions

In the majority of movements, one attachment of a muscle remains relatively stationary while the attachment at the other end moves. The more stationary attachment is called the *origin* of the muscle, and the other attachment, the *insertion*. A spring that closes a gate could be said to have its origin on the gatepost and its insertion on the gate itself.

In the body, however, the attachment arrangement is rarely so clear-cut, because, depending on the activity one is engaged in, the fixed and movable ends of the muscle may be reversed. For example, muscles that attach the upper limb to the chest normally move the arm relative to the trunk, which means their origins are on the trunk and their insertions are on the upper limb. However, in climbing, the arms are fixed, while the trunk moves as it is pulled up to the fixed limbs. In this type of situation, where the insertion is fixed and the origin moves, the muscle is said to perform a *reversed action*. Because there are so many situations where muscles are working with a reversed action, it is sometimes less confusing to simply speak of attachments, without reference to origin and insertion.

In practice, a muscle attachment that lies more proximally (more toward the trunk or on the trunk) is usually referred to as the *origin*. An attachment that lies more distally (away from the attached end of a limb or away from the trunk) is referred to as the *insertion*.

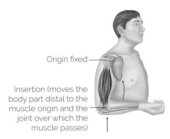

Origin fixed

Insertion (moves the body part distal to the muscle origin and the joint over which the muscle passes)

Figure 7.13: A muscle working with its origin fixed and its insertion moving.

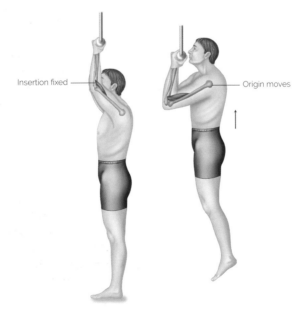

Insertion fixed

Origin moves

Figure 7.14: Climbing: the muscles are working with the insertion fixed and the origin moving (reversed action).

Group Action of Muscles

Muscles work together or in opposition in order to achieve a wide variety of movements; therefore, whatever one muscle can do, there is another muscle that can undo it. Muscles may also be required to provide additional support or stability to enable certain movements to occur elsewhere.

Muscles are classified into four functional groups:

- Prime mover, or agonist
- Antagonist
- Synergist
- Fixator

Prime Mover, or Agonist

A *prime mover* (also called an *agonist*) is a muscle that contracts to produce a specific movement. An example is the biceps brachii, which is the prime mover in elbow flexion. Other muscles may assist the prime mover in providing the same movement, albeit with less effect: such muscles are called *assistant* or *secondary movers*. For example, the brachialis assists the biceps brachii in flexing the elbow, and is therefore a secondary mover.

Antagonist

The muscle on the opposite side of a joint to the prime mover, and which must relax to allow the prime mover to contract, is called an *antagonist*. For example, when the biceps brachii on the front of the arm contracts to flex the elbow, the triceps brachii on the back of the arm must relax to allow this movement to occur. When the movement is reversed (i.e. the elbow is extended), the triceps brachii becomes the prime mover and the biceps brachii assumes the role of antagonist.

Synergist

Synergists prevent any unwanted movements that might occur as the prime mover contracts. This is especially important where a prime mover crosses two joints, because when it contracts it will cause movement at both joints, unless other muscles act to stabilize one of the joints. For example, the muscles that flex the fingers not only cross the finger joints, but also cross the wrist joint, potentially causing movement at both joints. However, because you have other muscles acting synergistically to stabilize the wrist joint, you are able to flex the fingers into a fist without also flexing the wrist at the same time.

A prime mover may have more than one action, and so synergists also act to eliminate the unwanted movements. For example, the biceps brachii will flex the elbow, but its line of pull will also supinate the forearm (twist the forearm, as in tightening a screw). If you want flexion to occur without supination, other muscles must contract to prevent this supination. In this context, such synergists are sometimes called *neutralizers*.

Fixator

A synergist is more specifically referred to as a *fixator* or *stabilizer* when it immobilizes the bone from which the prime mover takes origin, thus providing a stable base for the action of the prime mover. The muscles that stabilize (fix) the scapula during movements of the upper limb are good examples. The sit-up exercise is another good example. The abdominal muscles attach to both the ribcage and the pelvis; when they contract to enable you to perform a sit-up, the hip flexors will contract synergistically as fixators to prevent the abdominals tilting the pelvis, thereby enabling the upper body to curl forward as the pelvis remains stationary.

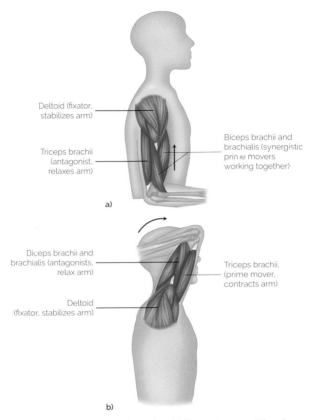

Figure 7.15: Group action of muscles: (a) flexing the arm at the elbow; (b) extending the arm at the elbow (showing reversed roles of prime mover and antagonist).

Leverage

In classical biomechanics, the bones, joints, and muscles together form a system of levers in the body to optimize the relative strength, range, and speed required of any given movement. The joints act as fulcrums, the muscles apply the effort and the bones bear the weight of the body part to be moved.

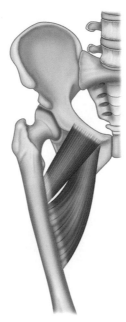

Figure 7.16: The pectineus is attached closer to the axis of movement than the adductor longus. Therefore, the pectineus is the weaker adductor of the hip, but is able to produce a greater movement of the lower limb per centimeter of contraction.

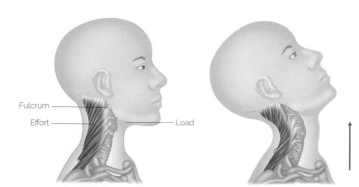

Figure 7.17: (continued).

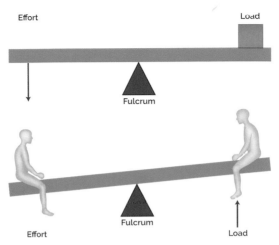

Figure 7.17: First-class lever: the relative position of the components is load–fulcrum–effort. Examples are a seesaw and a pair of scissors. In the body, an example is the ability to extend the head and neck: here the facial structures are the load, the atlanto-occipital joint is the fulcrum, and the posterior neck muscles provide the effort.

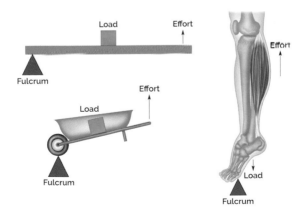

Figure 7.18: Second-class lever: the relative position of components is fulcrum–load–effort. The best example is a wheelbarrow. In the body, an example is the ability to raise the heels off the ground in standing: here the ball of the foot is the fulcrum, the body weight is the load, and the calf muscles provide the effort. With second-class levers, speed and range of movement are sacrificed for strength.

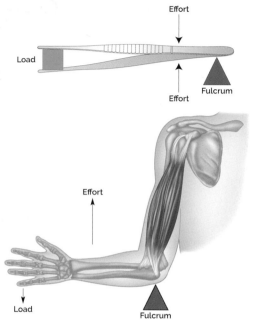

Figure 7.19: Third-class lever: the relative position of components is load–effort–fulcrum. A pair of tweezers is an example of this. In the body, most skeletal muscles act in this way. An example is flexing the forearm: here an object held in the hand is the load, the biceps brachii provides the effort, and the elbow joint is the fulcrum. With third-class levers, strength is sacrificed for speed and range of movement.

A muscle attached close to the fulcrum will be weaker than it would be if it were attached further away. However, it is able to produce a greater range and speed of movement, because the length of the lever amplifies the distance travelled by its movable attachment. Figure 7.16 illustrates this principle in relation to the adductors of the hip joint. The muscle so positioned to move the greater load (in this case the adductor longus) is said to have a *mechanical advantage*. The muscle attached close to the fulcrum is said to operate at a *mechanical disadvantage*, and has a weaker action.

Figures 7.17–7.19 illustrate the differences in first-, second-, and third-class levers, by means of examples in the human body.

Muscle Factors That Limit Skeletal Movement

The inability of a muscle to contract or lengthen beyond a certain point can cause some practical hindrances to bodily movement: these are outlined below.

Passive Insufficiency

Muscles that span two joints are called *biarticular*. These muscles may be unable to "pay out" sufficiently to allow full movement of both joints simultaneously, unless the muscle has been trained to relax. For example, most people need to bend their knees in order to touch their toes; this is because the hamstrings (which span the hip and knee joints) cannot lengthen enough to allow full flexion at the hip joint without also pulling the knee joint into flexion. For the same reason,

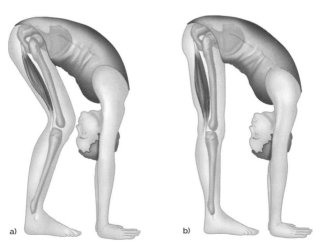

a)   b)

Figure 7.20: Passive insufficiency example 1: (a) having to bend the knees to the touch toes means there is passive insufficiency of the hamstrings; (b) being able to touch the toes with the knees straight means that there is much less passive insufficiency of the hamstrings.

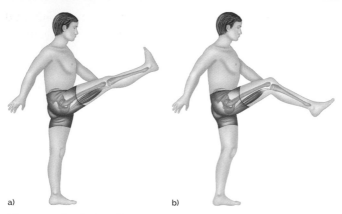

a)   b)

Figure 7.21: Passive insufficiency example 2: (a) a high kick with knee straight is possible only if the hamstrings have been trained to overcome their passive insufficiency; (b) for most people, an attempt to perform a high kick will be restricted by hamstring passive insufficiency, causing the knee to bend.

it is easier to pull your thigh to your chest if your knee is bent than it is with your knee straight. This limitation is called *passive insufficiency*. Passive insufficiency is therefore the inability of a muscle to lengthen by more than a fixed percentage of its length.

Active Insufficiency

Active insufficiency is the opposite of passive insufficiency. Whereas passive insufficiency results from the inability of a muscle to lengthen by more than a fixed percentage of its length, active insufficiency results from the inability of a muscle to contract by more than a fixed amount. For example, most people can flex their knee to bring their heel close to their buttock when their hip is flexed, because the upper part of the hamstrings is lengthened and the lower part is shortened. However, one is normally unable to fully flex the knee when the hip is extended; this is because with the hip extended, the hamstrings are already shortened, meaning that there is insufficient "shortening" potential remaining in the hamstrings to then fully flex the knee.

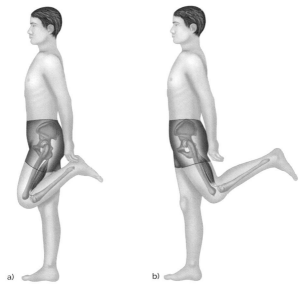

Figure 7.22: Active insufficiency: (a) with the hip flexed, the hamstrings are stretched at the hip, enabling their contraction to fully flex the knee; (b) with the hip extended, the shortened hamstrings are unable to contract any further to fully flex the knee.

Concurrent Movement

If extension of the hip is required at the same time as extension of the knee, as in the push-off from the ground in running, the phenomenon known as *concurrent movement* applies, and proves very useful. To grasp the concept of concurrent movement, first remember that when the hamstrings contract, they are able to both extend the hip joint and flex the knee joint, either singly or simultaneously. In analyzing the example of running in more detail, we can therefore observe the following:

- As the foot pushes against the ground, the hamstrings contract to extend the hip.
- Meanwhile, the fixators prevent the hamstrings from flexing the knee.

- Consequently, the hamstrings are shortened only at their upper end (origin), but remain lengthened at their lower end (insertion).
- The antagonist to the action of the hamstrings in extending the hip is the rectus femoris, which relaxes because of reciprocal inhibition to allow the hamstrings to contract.
- When the hip is fully extended, the already stretched rectus femoris is unable to lengthen further, causing it to pull the knee into extension.
- Thus, the rectus femoris is lengthened at its upper end and shortened at its lower end.

Concurrent movement therefore avoids passive and active insufficiency of the hamstrings and rectus femoris by neither shortening nor stretching both ends of either muscle; rather, one end lengthens as the other shortens, and vice versa in the other muscle. Figure 7.23 illustrates this concept.

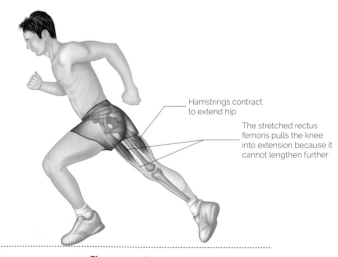

Hamstrings contract to extend hip

The stretched rectus femoris pulls the knee into extension because it cannot lengthen further

Figure 7.23: Concurrent movement.

Countercurrent Movement

If flexion of the hip is required to occur at the same time as extending the knee, as in kicking a ball, a *countercurrent movement* occurs. In analyzing the example of kicking in more detail, we therefore observe the following:

- In the action of kicking a ball, the rectus femoris acts as a prime mover to flex the hip and extend the knee.
- Thus, both the upper and lower portions of rectus femoris are shortened.
- The hamstrings relax because of reciprocal inhibition, so that they can extend at both ends and allow the kick to occur.
- The rectus femoris relaxes once the movement has been made, but the momentum of the movement is still propelling the leg forward.
- At this stage, the hamstrings contract to act as a "brake" for the leg, as it flies forward.

Countercurrent movements therefore prevent injury by ensuring the antagonist relaxes first, then contracts at the right time to prevent the forces of momentum from overstretching the muscles and ligaments. So-called *ballistic* movements rely on this principle, but are often done so forcefully that the power of momentum is greater than the ability of the antagonist to "brake" that momentum. In such instances, muscle and ligament damage often occurs.

Core Stability

Core stability could be described as 'the capacity for the trunk to support, control and withstand the forces acting upon it, so that the body structures can perform in their safest, strongest and most efficient positions' (Elphinston, 2015).

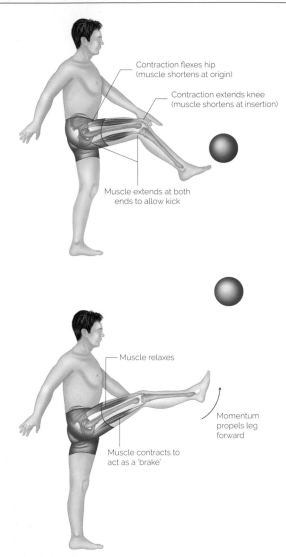

Contraction flexes hip
(muscle shortens at origin)

Contraction extends knee
(muscle shortens at insertion)

Muscle extends at both
ends to allow kick

Muscle relaxes

Momentum
propels leg
forward

Muscle contracts to
act as a 'brake'

Figure 7.24: Countercurrent movement.

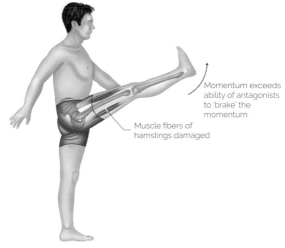

Momentum exceeds ability of antagonists to 'brake' the momentum

Muscle fibers of hamstings damaged

Figure 7.25: Damage that can be caused by an overzealous ballistic stretch.

This capacity to manage forces by the central body is one of many factors involved in the overall functional stability of the body in motion. The muscles associated with the "core" must be able to function in a variety of ways to meet our functional needs. The task may require endurance to support the trunk and pelvis against the strong muscular action of the legs for example, or rapid responsiveness to withstand a sudden blow like a rugby tackle.

Stabilisation of the trunk is produced by the interaction of multiple muscles. Some, like the multifidus muscles directly adjacent to their corresponding spinal segment, are short fibred and located close to their area of influence, and as such have relatively minimal movement effect. Others such as the erector spinae, can exert force over multiple spinal segments. Some, such as latissimus dorsi, gluteus maximus and hamstrings, are seemingly distant from the spine itself, yet provide support via their fascial connections.

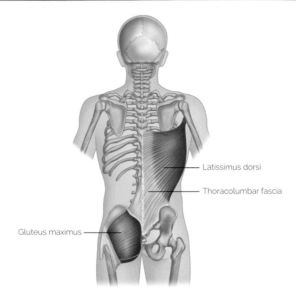

Figure 7.26: The posterior oblique myofascial sling: latissimus dorsi, gluteus maximus and thoracolumbar fascia.

Latissimus dorsi

Thoracolumbar fascia

Gluteus maximus

The muscles most commonly associated with core stability are the abdominals. The more superficial of these, such as the external obliques, create torque, and as such are forceful muscles capable of being trained for effective strength. The deepest of the abdominal group, the transversus abdominis, works at a low level of recruitment, functioning as a transmitter of force between lower and upper body (Allison et al 2008), a connector of the anterior to posterior trunk via its attachments into the thoracolumbar fascia, and an important component of the body's natural stabilising mechanism, intra–abdominal pressure (Hodges et al 2005).

Early research by Hodges and Richardson (1997) on the behaviour of transversus abdominis generated widespread change in rehabilitation approaches, especially to low back pain. Their findings indicated that this muscle acts in a

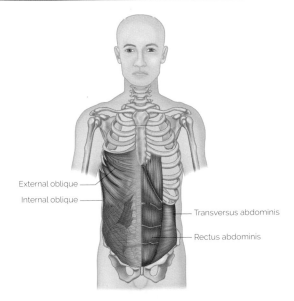

Figure 7.27: The abdominal muscles are those most commonly associated with core stability.

feedforward mechanism, i.e. it responds to the impulse to move, in order to contribute to the body's overall preparation to manage the force created by movement. Its role as one of many muscles inserting into the thoracolumbar fascia further strengthened the case for training it, especially when it was found that its function was altered in the presence of low back pain (Hodges 2001).

However, after many years of research, the performance of key muscles such as transversus abdominis has not been convincingly associated with improved outcomes for low back pain (Allison et al. 2012; Lim et al. 2011), nor has core stability training in general been shown to be more effective than general exercise for this patient group (Koumantakis et al. 2005). Core stability training is therefore now considered within the broader context of the total movement function of an individual.

In the case of intra-abdominal pressure, the diaphragm and pelvic floor form a roof and floor respectively to an imaginary inflatable canister sitting in front of the spine, with the transversus abdominis and posterior fibres of internal obliques primarily wrapping around to form its walls. The muscles are finely synchronised to modify the amount of pressure in this canister depending upon what we are doing, and this in turn helps to stabilise the spine from within. It is coordination of all of the components that creates the resulting spinal support, rather than an emphasis on isolated muscle behaviour.

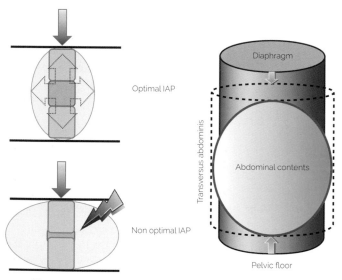

Figure 7.28: Generation of the optimum levels of pressure inside the thoracic and abdominal cavities.

Structurally, those muscles which are predominantly "mover" muscles and which can exert high force tend to be relatively superficial to take advantage of longer lines of pull on the limbs. Those muscles, which work predominantly as controllers of joint motion, tend to be situated more deeply to support and maintain joint position throughout a movement.

In the case of dysfunction or inhibition of these muscles, the more superficial muscles may become facilitated to provide a form of compensatory stability. When this happens the individual experiences a loss of joint range of motion, which stretching does not improve. Training must therefore address the balance and coordination of muscles with good quality movement in order to restore equilibrium.

Where once there was an emphasis on training a constant protective bracing behaviour from the central core muscles, now it has been identified that individuals with low back pain can tend to exhibit too much activity rather than too little (Brumagne et al. 2008), or an inability to release trunk muscles adequately to respond to sudden movements (Cholewicki 2005). This response can be both physiological and behavioural, emphasising the need to train in such a way that incorporates smooth movement, confidence, body awareness and coordination.

Pelvic control is as critical to maintenance of trunk stability as the muscles of the trunk itself. Anatomical bridges between the pelvis and trunk such as psoas are strongly implicated in this, highlighting the intimate connection between hip, pelvis and trunk posture. Pelvic position such as anterior, posterior or lateral tilt will directly influence trunk position, and the trunk musculature must respond accordingly. Pelvic posture, as well as strength and timing of the pelvic musculature such as the gluteal group will therefore be an essential component of any trunk rehabilitation programme.

References

Allison, G.T. Morris, S.L. & Lay, B. 2008. Feedforward responses of transversus abdominis are directionally specific and act asymmetrically: implications for core stability theories. *J Orthop Sports Phys Ther* 38(5), 228–237.

Allison, G.T. 2012. Abdominal muscle feedforward activation in patients with chronic low back pain is largely unaffected by 8 weeks of core stability training. *J Physiother* 58(3), 200.

Borghuis, J. Hof, A.L. & Lemmink, K.A. 2008. The importance of sensory-motor control in providing core stability: implications for measurement and training. *Sports Med* 38(11), 893–916.

Brumagne, S. Janssens, L. Knapen, S. Claeys, K. & Suuden-Johanson, E. 2008. Persons with recurrent low back pain exhibit a rigid postural control strategy. *Eur Spine J* 17(9), 1177–1184. doi: 10.1007/s00586-008-0709-7. Epub Jul 2.

Cholewicki, J. Silfies, S.P. Shah, R.A. Greene, H.S. Reeves, N.P. Alvi, K. & Goldberg, B. 2005. Delayed trunk muscle reflex responses increase the risk of low back injuries. *Spine* 30(23), 2614–2620.

Elphinston, J. 2013. *Stability, Sport and Performance Movement: Practical Biomechanics and Systematic Training for Movement Efficacy and Injury Prevention, Second Edition.* Lotus Publishing, Chichester.

Hodges, P.W. Eriksson, A.E. Shirley, D. & Gandevia, S.C. 2005. Intra-abdominal pressure increases stiffness of the lumbar spine. *J Biomech* 38(9), 1873–1880.

Hodges, P. 2001. Changes in motor planning of feedforward postural responses of the trunk muscles in low back pain. *Experimental Brain Research* 141(2), 261–266.

Hodges, P. & Richardson, C.A. 1997. Feedforward contraction of transversus abdominis is not influenced by the direction of arm movement. *Experimental Brain Research* 114(2), 362–370.

Koumantakis, G.A. Watson, P.J. & Oldham, J.A. 2005. Trunk muscle stabilization training plus general exercise versus general exercise only: randomized controlled trial of patients with recurrent low back pain. *Phys Ther* 85(3), 209–225.

Lim, E.C. Poh, R.L. Low, A.Y. & Wong, W.P. 2011. Effects of Pilates-based exercises on pain and disability in individuals with persistent nonspecific low back pain: a systematic review with meta-analysis. *J Orthop Sports Phys Ther* 41(2), 70–80.

CHAPTER 8

Fascia and the Anatomy Trains Myofascial Meridians

Thomas Myers

The Anatomy Trains *myofascial meridians* presents a map of how tracks of fascial fabric wind longitudinally through series of muscles. This new approach to structural patterning has far-reaching implications for effective movement training and manual therapy treatment at some distance from the site of dysfunction or pain, especially for long-standing postural imbalances, unsound body usage, and sequelae from surgery, injury or insult. These ideas are unfolded in greater detail in the book *Anatomy Trains* (Elsevier, 2014), and at AnatomyTrains.com.

Practical Holism

Humans are not assembled out of parts like a car or a computer. The heart is a pump; the brain is a computer, the kidney is a filter – all these 'Body as machine' images are useful metaphors, but like any poetic trope, they do not tell the whole story. In our modern perception of human movement anatomy, however, we are in danger of making this mechanical metaphor into the be all and end all. In actual fact, our bodies are conceived as a whole, and grow, live, and finally die as a whole – but our mind is a knife (Figure 8.1).

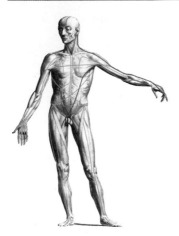

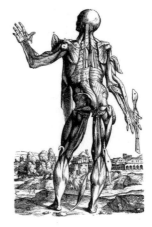

Figure 8.1: The Anatomy Trains map of myofascial connections.

Figure 8.2: Vesalius's woodcuts from 1548 show both the origami layering and the directional 'grain' of the myofascial system.

The tool of choice for anatomy was a blade. From the flint cleaver to the laser scalpel, the animal and human body has been divided along finer and finer lines. Cartesian dualism described the body as a 'soft machine', and students of anatomy and physiology used reductionistic mechanism to go about explaining the role of each identifiable part. Newton's laws further cemented our place within the mechanical universe. What were glorious and liberating ideas in their own time, however, have become imprisoning, restrictive concepts in ours (Figure 8.2).

How do our 'parts' really arise? Like a plant, from a 'seed'. The human body stems from a single fertilized human ovum, which proliferates wildly. The daughter cells then specialize as each tissue cell exaggerates some function of the ovum and cells in general – e.g. a muscle specializes in contraction, a neuron in conduction, epithelia in secretion, etc. – and conversely other functions diminish. A nerve cell conducts extraordinarily well, but as a result of that specialization cannot easily reproduce itself. Epithelia do very well at creating enzymes, but lose the ability to significantly contract. Yet each cell still partakes of the unique individual whole in its constant communication with its neighbors, near and far, and in the similarities

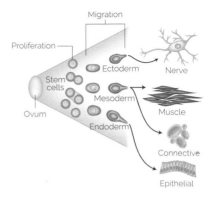

Figure 8.3: From the generalized ovum, cells proliferate, migrate, and differentiate into functionally specialized tissues.

of chemical structure, from glucose as a universal fuel right on down to the spooled helix of DNA (Figure 8.3).

Before specific cuts are made, what we are pleased to call 'the brain' never exists as an entity separate from its connective tissue surroundings, its blood supply, and the peripheral and autonomic nerves that extend the brain throughout the body. The 'biceps brachii' can only exist as a separate structure with a knife's intervention to divide not only its ends but other various attachments, its connections with surrounding myofascial units such as the brachialis, as well as its nerve and blood supply, without which it simply could not function. The idea that there are separate parts – a liver, a brain, a biceps – may be the way that we think, but it is not the way physiology 'thinks'.

The Single Muscle Theory

This image of 'separate' muscles – muscles as individual actors on bones via tendons over joints – leads to the prevalent method of analyzing muscle action, which is employed frequently (and to good purpose) throughout anatomy atlases: "Imagine that the skeleton were denuded of all but the given muscle; what would that single muscle do to the skeleton acting on its own?" Call this the 'single muscle' theory.

In this single muscle theory, the biceps gets defined as a radio-ulnar supinator, an elbow flexor, and a weak flexor of the shoulder (Figure 8.4a). In the Anatomy Trains view, additional information is added to this: "The biceps brachii is an element in a continuous fascial plane or myofascial meridian which runs from the outside of the thumb to the fourth rib and beyond." The second statement does not negate the first, but it adds a context for understanding the biceps' role in stabilizing the thumb (down the myofascial line), and keeping the chest open and the breath full (up the line) (Figure 8.4b).

This 'body as assembled machine' idea is so pervasive – and as in this book, the maps based on this perspective are so understandable and useful – that it is difficult to think outside its parameters. Thinking in 'wholes', attractive as it is to

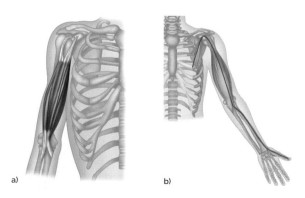

a)   b)

Figure 8.4: (a) the biceps brachii considered as a separate muscle, (b) the biceps brachii is also part of a longitudinal myofascial continuity.

contemporary holistic therapists, simply has yet to lead to useful maps to bridge traditional Western anatomy with, say, the Asian maps for acupuncture, or reflex maps of the hands, feet, eye, or ear.

The 'everything is connected to everything else' philosophy expounded in our opening paragraphs, while actually technically accurate, leaves the practitioner adrift in this sea of connections, unsure as to whether that frozen shoulder will respond to work in the elbow, the contralateral hip, or to a reflex point on the ipsilateral foot. While any of these might work, useful maps are necessary to organize our therapeutic choices into something better than a guess.

In short, we know the body is interconnected on many levels, but we need better treatment strategies than 'press and pray' – or, for the movement therapist, 'stretch and pray'. What can we learn when we shift from a 'symptom-oriented' view of the body to a 'system-oriented' one?

This myofascial meridians concept provides such a map of the structural body, providing a practical transition between the individual 'parts' that the authors have so brilliantly catalogued herein, and the 'whole' of a human being, a gestalt of physics, physiology, stored experience, and current awareness which defies mapping. We hope that you will agree that Anatomy Trains, this intermediate map of the body's locomotor fabric, opens up new avenues of treatment consideration, particularly for stubborn chronic conditions and global postural effects.

Whole Body Communicating Networks

Central to this new 'Anatomy Trains' is the functional unity of the connective tissue system. There are exactly three networks within the body which, magically extracted intact, would show us the shape of the whole body, inside and out: the neural net, the vascular system, and the extracellular fibrous web (Extra-Cellular Matrix or ECM) created by the connective tissue cells (Figure 8.5).

All three networks communicate throughout the body. The nerves carry sensory data to the center, which constructs a second-to-second picture of the world, and then carry signals out to the muscles and glands, at speeds between 7 and 170 miles per hour. The fluid systems circulate chemistry around the body every few minutes, though many chemical rhythms fluctuate in hourly, daily, or as women know, monthly cycles.

The ECM communicates mechanical information – tension and compression – via fasciae, tendon, ligament, and bone – and, as we know now, even the softest, loosest tissue like fat can be a very effective force transmitter. This information is a vibration that travels at the speed of sound, about 700 mph – slower than light, but minimally 4× faster than the nervous system. The speed of response – plastic deformation and compensation in the connective tissue system – is however measured in hours to weeks, months to years. Thus the fibrous system is both the fastest (in communicating) and the slowest (in responding) of the three.

Unlike the neural and vascular systems, this connective tissue net is not yet well mapped,

Figure 8.5: The Anatomy Trains map (posterior view).

because it is considered to be the 'dead' material that we need to remove to see the 'interesting' neural, vascular, muscular, and other local systems. Because the connective tissue provides the divisions along which the scalpel runs to parse out other systems, the connective tissue has also been studied less as a system than other more familiar systems.

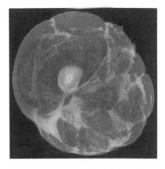

Figure 8.6: A section of the thigh, viewed from above, with all other tissues but the connective tissues removed. Courtesy of Jeff Linn, derived from the Visible Human Data Project.

So, as a thought experiment: what if, instead of dividing the body into individual identifiable structures, we were to dip it into a solvent that shampooed away all the cellular material but left the entire extracellular matrix intact (Figure 8.6)?

The Connective Tissue System

This system of the connective tissue matrix can be seen as our 'organ of form'. From the moment of the first division of the ovum, the ECM of the connective tissue exists as a secreted glycoaminoglycans (mucous) gel that acts to glue the cells together. Around the end of the second week of embryological development, the first fibrous version of this net appears, a web of fine reticulin spun by specialized mesodermal cells on either side of the developing notochord (spine). This net is the origin of our fascial web – our 'metamembrane' that weaves us together – the singular container that shapes our form and directs the flow of all our biochemical processes (Figure 8.7).

The ability of the connective tissue cells to alter and mix the three elements of the intercellular space – the water, the

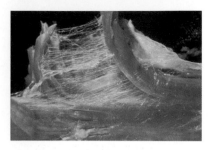

Figure 8.7: A magnification of myofascial tissue – individual muscle fibers within the cotton floss of the endomysial fascia. Photograph courtesy of Ron Thompson.

fibers, and the gluey ground substance gel of glycoaminoglycans – produces on demand the wide spectrum of familiar building materials in the body – bone, cartilage, sacs and bags, heart valves, eye cornea, tooth dentin, ligament, tendon, areolar and adipose networks – all the varieties of biological fabric. The body's joints, the 'organ system of movement', are almost entirely composed of ECM constructed by fibroblast cells and their cousins. This ECM, taken as a whole, not only unites the various elements of the body; in large part it unites the many branches of medicine.

Tensegrity

Another holistic image is necessary to jump out of this 'machine made out of parts' image so ingrained in our systems – tensegrity geometry. The normal geometric picture of our anatomy is that the skeleton is a continuous compression framework, like a crane or a stack of blocks, and the muscles hang from it like the cables. This leads to the single muscle theory again – the skeleton is stable but moveable, and we parse out what each muscle does to that framework on its own, adding them together to analyze the movement. A little thought, however, soon puts this idea out to pasture. Take the muscles away, and the skeleton is anything but stable; take all the soft-tissue away and the bones would clatter to the floor, as they do not interlock or stack in any kind of stable way.

If we can get away from the idea that bones are like girders and muscles are the cables that move the girders, we are led to a class of structures called 'tensegrity' (the integrity lies in the balance of tension) (Figure 8.8). Originated by Kenneth Snelson and developed by Buckminster Fuller, tensegrity geometry more closely approximates the body as we live and feel it than does the old 'crane' model. In the dance of stability and mobility that is a human moving, the bones and cartilage are clearly compression-resisting struts that push outward against the myofascial net. The net, in turn, is always tensional, always trying to pull inward toward the center. Both elements – centrifugal and centripetal – are necessary for stability, and both contribute to practical mobility.

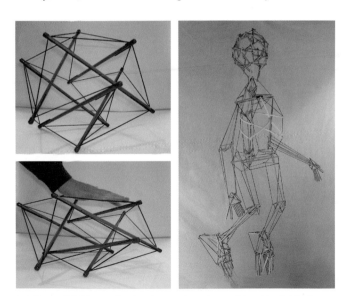

Figure 8.8: Tensegrity structures, when stressed, tend to distribute rather than concentrate strain. The body does the same, with the result that local injuries soon become global strain patterns. For more on tensegrity:kennethsnelson.net, biotensegrity.com http://web1.tch.harvard.edu/research/ingber/Tensegrity.html. Main photograph courtesy of Tom Flemons, intensiondesigns.com.

Tensegrity is the only scalable geometric model that could have served organisms from the single-celled that predominated for over half life's history to the trillion-celled communities that crawled out of the water into gravity and still walk the Earth today. One of the weaknesses of our old biomechanical model is that it does not work on the molecular or tissue scale; tensegrity works at all scales of magnitude.

In this new orthopedic model, the bony struts 'float' within the sea of tension provided by the soft tissues. Thus the position of the bones is thus dependent on the tensional balance among these soft-tissue elements. This model is of great importance in seeing the larger potential of soft-tissue approaches to structure, e.g. yoga, Pilates, personal training, and bodywork. In reality, bony position and posture is far more dependent on soft-tissue balance than on any high-velocity thrusting of bones back into 'alignment'.

In this view, much expanded in our other writings, the Anatomy Trains Myofascial Meridians map the global lines of tension that traverse the entire body's muscular surface, acting to keep the skeleton in shape, guide the available tracks for movement, and coordinate global postural stability with the local mobility needed to get the groceries out of the car. Research supports the idea of tensegrity geometry ruling mechanical transmission from the cellular level on up, and macro-level models, such as the one pictured here, are becoming more anatomically accurate.

The Anatomy Trains

Let us now step down to intermediate level, somewhere between these overarching global considerations and the useful detailed anatomy that comprises the rest of this book. The concept is very simple: if we follow the grain of the fascial fabric, we can see where muscles are linked up longitudinally. When this is done, there are twelve or so major myofascial meridians that appear, forming clear lines, or tracks, that traverse the body.

We can construct twelve myofascial meridians in common use in human stance and movement:

- Superficial Front Line
- Superficial Back Line
- Lateral Line
- Spiral Line
- Arm Lines (four)
- Functional Lines (three – front, back and ipsilateral)
- Deep Front Line

The first three lines are termed the 'cardinal' lines, in that they run more or less straight up and down the body in the four cardinal directions – front, back, and left and right sides.

Superficial Front Line
The Superficial Front Line (SFL) runs on both the right and left sides of the body from the top of the foot to the skull, including the muscles and associated fascia of the anterior compartment of the shin, the quadriceps, the rectus abdominis, sternal fascia, and sternocleidomastoid muscle up onto the galea aponeurotica of the skull. In terms of muscles and

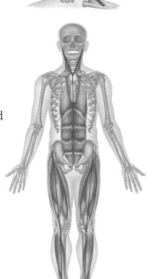

Figure 8.9: The Superficial Front Line (SFL).

tensional forces, the SFL runs in two pieces – toes to pelvis, and pelvis to head, which function as one piece when the hip is extended, as in standing (Figure 8.9).

In the SFL, fast-twitch muscle fibers predominate. The SFL functions in movement to flex the trunk and hips, to extend the knee, and to dorsiflex the foot. In standing posture, the SFL flexes the lower neck but hyperextends the upper neck.

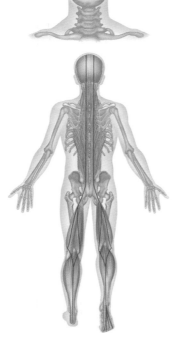

Posturally, the SFL also maintains knee and ankle extension, protects the soft organs of the ventral cavity, and provides tensile support to lift those parts of the skeleton which extend forward of the gravity line – the pubes, the ribcage, and the face. And, of course, it provides a balance to the pull of the Superficial Back Line.

A common human response to shock or attack, the startle response, can be seen as a shortening of the SFL. Chronic contraction of this line – common after trauma, for example – creates many postural pain patterns, pulling the front down and straining the back.

Superficial Back Line

The Superficial Back Line (SBL) runs from the bottom

Figure 8.10: The Superficial Back Line (SBL).

of the toes around the heel and up the back of the body, crossing over the head to its terminus at the frontal ridge at the eyebrows. Like the SFL, it also has two pieces, toes to knees and knees to head, which function as one when the knee is extended (as in most forward bend yoga asanas, for instance). It includes the plantar tissues, the triceps surae, the hamstrings and sacrotuberous ligament, the erector spinae, and the epicranial fascia.

The SBL functions in movement to extend the spine and hips, but to flex the knee and ankle. The SBL lifts the baby's eyes from primary embryological flexion, progressively lifting the body to standing (Figure 8.10).

Posturally, the SBL maintains the body in standing, spanning the series of primary and secondary curves of the skeleton (including the cranium and heel in the catalogue of primary curves, and knee and foot arches in the list of secondary curves). This results in a more densely fascial line than the SFL, with strong bands in the legs and spine, and a predominance of slow-twitch fibers in the muscular portion.

Lateral Line

The Lateral Line (LL) traverses each side of the body from the medial and lateral midpoints of the foot around the fibular malleolus and up the lateral aspects of the leg and thigh, passing along the trunk in a woven pattern that extends to the skull's mastoid process (Figure 8.11).

In movement, the LL creates lateral flexion in the spine, abduction at the hip, and eversion at the foot, and also operates as an

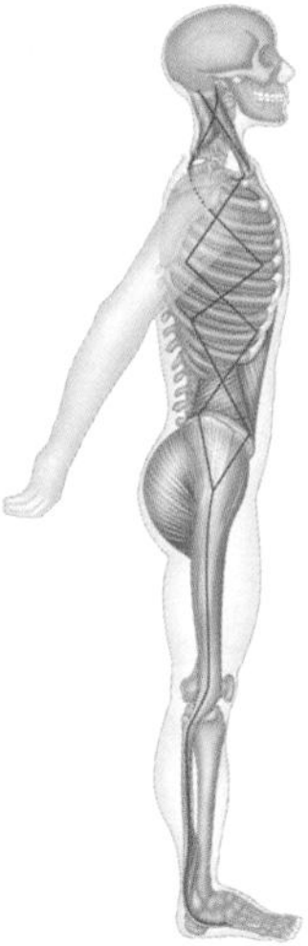

Figure 8.11: The Lateral Line (LL).

adjustable 'brake' for lateral and rotational movements of the trunk. The LL acts posturally like tent guy-wires to balance the left and right sides of the body. Also, the LL contains more than creates movement in the human, directing the flexion-extension that characterizes our direction through the world, restricting side-to-side movement that would otherwise be energetically wasteful.

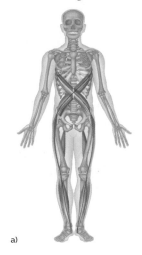

a)

Spiral Line

The Spiral Line (SL) winds through the three cardinal lines, looping around the trunk in a helix, with another loop in the legs from hip to arch and back again. It joins one side of the skull across the midline of the back to the opposite shoulder, and then across the front of the torso to the same side hip, knee and foot arch returning up the back of the body to the head (Figure 8.12).

In movement, the SL creates and mediates rotations in the body. The SL interacts with the other cardinal lines in a multiplicity of functions. In posture, the SL wraps the torso in a double helix that helps to maintain spinal length and balance in all planes. The SL connects the foot arches with tracking of the knee and

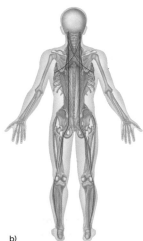

b)

Figure 8.12: The Spiral Line (SL); (a) anterior view, (b) posterior view.

pelvic position. The SL often compensates for deeper rotations in the spine or pelvic core.

Arm Lines

- Superficial Front Arm Line
- Superficial Back Arm Line
- Deep Front Arm Line
- Deep Back Arm Line

The four Arm Lines run from the front and back of the axial torso to the tips of the fingers. They are named for their planar relation in the composition of the shoulder, and roughly parallel the four lines in the leg. These lines connect seamlessly into the other lines particularly the Lateral, Functional, Spiral, and Superficial Front Lines (Figure 8.13).

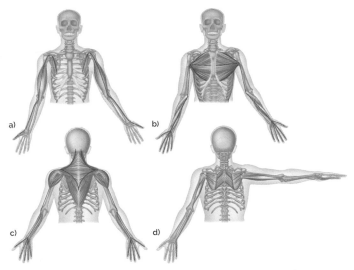

Figure 8.13: The four Arm Lines; (a) Superficial Front Arm Line, (b) Deep Front Arm Line, (c) Superficial Back Arm Line, (d) Deep Back Arm Line.

In movement the arm lines place the hand in appropriate positions for the task before us – examining, manipulating, or responding to the environment. The Arm Lines act across ten or more levels of joints in the arm and shoulder to bring things to us or to push them away, to push, pull, or stabilize our own bodies, or simply to hold some part of the world still for our perusal or modification. The Arm Lines affect posture indirectly, since they are not part of the structural column. Given the weight of the shoulders and arms, however, displacement of the shoulders in stillness or in movement will affect other lines – and notably the breathing pattern. Conversely, structural displacement of the trunk in turn affects the arms' effectiveness in specific tasks and may predispose them to injury.

Beyond the straightforward progression of the meridians from the trunk to the four corners of the hands, there are many 'crossover' muscles that link these lines to ether, providing additional support and stability for the extra mobility the arms have relative to the legs.

Functional Lines
- Front Functional Line
- Back Functional Line
- Ipsilateral Functional Line

The Front and Back Functional Lines join the contralateral girdles across the front and back of the body, running from one humerus to the opposite femur and vice versa (Figure 8.14). The Ipsilateral Functional Line joins the humerus to the inner knee on the same side.

The Functional Lines are used in innumerable movements, from walking to the most extreme sports. They act to extend the levers of the arms to the opposite leg as in a kayak paddle, a baseball throw or a cricket pitch (or vice versa in the case of a football kick). Like the Spiral Line, the Functional Lines are helical, and thus help create strong rotational movement. Their postural function is minimal.

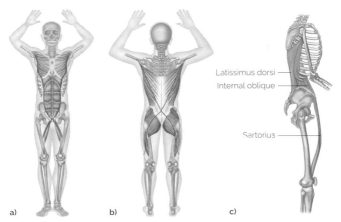

Latissimus dorsi

Internal oblique

Sartorius

a) b) c)

Figure 8.14: The Functional Lines. (a) Front Functional Line; (b) Back Functional Line; (c) Ipsilateral Functional Line.

Deep Front Line

The Deep Front Line (DFL) forms a complex core volume from the inner arch of the foot, up the inseam of the leg, into the pelvis and up the front of the spine to the bottom of the skull and the jaw. This 'core' line lies between the Front and Back Lines in the sagittal plane, between the two Lateral Lines coronally, and is wrapped circumferentially by the Spiral and Functional Lines. This line contains many of the more obscure supporting muscles of our anatomy, and because of its internal position has the greatest fascial density of any of the lines (Figure 8.15).

Structurally, this line has an intimate connection with the arches, the hip joint, lumbar support, and neck balance. Functionally, it connects the ebb and flow of breathing (dictated by the diaphragm) to the rhythm of walking

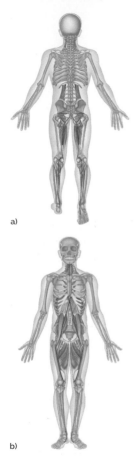

a)

b)

Figure 8.15: The Deep Front Line (DFL); (a) anterior view, (b) posterior view.

(organized by the psoas). In the trunk, the DFL is intimately linked with the autonomic ganglia, and thus uniquely involved in the sympathetic/parasympathetic balance between our neuro-motor 'chassis' and the ancient organs of cell-support in the ventral cavity.

The importance of the DFL to posture, movement, and attitude cannot be over-emphasized. A dimensional understanding of the DFL is necessary for successful application of nearly any method of manual or movement therapy. Because many of the movement functions of the DFL are redundant to the superficial lines, dysfunction within the DFL can be barely visible in the outset, but these dysfunctions will gradually lead to larger problems. Restoration of proper DFL functioning is by far the best preventive measure for structural and movement therapies.

In conclusion, let us remember that Anatomy Trains is simply another map, an alternative way of viewing the myofascial system. It is not the answer to all questions, but is does provide a basis for global postural treatment and functional linkages to understand the distributions/compensations that happen post-injury or because of habitual usage patterns. We welcome your joining our explorations into these concepts and protocols at AnatomyTrains.com.

Muscle Innervation Pathways

Cranial Nerves

The cranial nerves emerge directly from the brain or brainstem, whereas the spinal nerves emerge directly from the spinal cord. Cranial nerves are listed below, and those that supply the specific skeletal muscles discussed in this book are covered in more detail.

Cranial nerve I, the olfactory nerve, is responsible for carrying sensory information relating to the sense of smell. **Cranial nerve II**, the optic nerve, is responsible for conveying visual information from the retina to the brain.

Cranial nerve III, the oculomotor nerve, controls most of the movements of the eye (along with cranial nerves IV and VI) and innervates the levator palpebrae superioris. **Cranial nerve IV**, the trochlear nerve, is a motor nerve that innervates a single muscle, the superior oblique muscle of the eye (not covered in this book).

Cranial nerve V, the trigeminal nerve, is the largest of the cranial nerves and has three main divisions: ophthalmic (V_1), maxillary (V_2), and mandibular (V_3). The trigeminal nerve is responsible for sensation in the face and for functions such as biting and chewing. Both the **ophthalmic division** and the **maxillary division** are purely sensory, while the

mandibular division has both sensory and motor functions. The mandibular division innervates masseter, temporalis, pterygoids, mylohyoid, and digastric (anterior belly).

Cranial nerve VI, the abducent nerve, controls the movement of just one muscle, the lateral rectus muscle of the eye (not covered in this book).

From the pons of the midbrain, **cranial nerve VII**, the facial nerve, enters the temporal bone through the internal acoustic meatus, and then emerges through the stylomastoid foramen, where it branches into the **posterior auricular branch**. The five major branches—temporal, zygomatic, buccal, (marginal) mandibular, and cervical (remember the mnemonic "**T**o **Z**anzibar **B**y **M**otor **C**ar")—innervate the facial muscles as follows.

Temporal branches: frontalis, temporoparietalis, auricularis anterior and superior, orbicularis oculi (also innervated by the zygomatic branches), procerus, and corrugator supercilii. **Zygomatic branches:** orbicularis oculi (also innervated by the temporal branches) and zygomaticus major (also innervated by the buccal branches). **Buccal branches:** depressor septi nasi, orbicularis oris (also innervated by the mandibular branches), levator labii superioris, levator anguli oris, nasalis, zygomaticus major (also innervated by the zygomatic branches), zygomaticus minor, depressor anguli oris, risorius, and buccinator. **Mandibular branches:** orbicularis oris (also innervated by the buccal branches), depressor labii inferioris, depressor anguli oris (also innervated by the buccal branches), mentalis, and stylohyoid. **Cervical branches:** platysma. Furthermore, the posterior auricular branch subdivides into the **auricular branch**, which innervates the auricularis posterior, and the **occipital branch**, which innervates the occipitalis. The **digastric branch**, which arises close to the stylomastoid foramen, innervates digastric.

Cranial nerve VIII, the vestibulocochlear nerve (also known as the *auditory vestibular nerve*), transmits sound and equilibrium (balance) information from the inner ear to the brain.

Cranial nerve IX, the glossopharyngeal nerve, originates from the medulla oblongata and exits the skull through the jugular foramen. Its main function is sensory.

Cranial nerve X, the vagus nerve, supplies motor parasympathetic fibers to all organs except the adrenal glands.

Cranial nerve XI, the accessory nerve, is unique in that it is formed by both cranial and spinal components that combine and then diverge, with the cranial portion joining the vagus nerve (X), and the spinal portion descending to innervate sternocleidomastoid and trapezius.

Cranial nerve XII, the hypoglossal nerve, innervates muscles of the tongue, although geniohyoid is innervated by the fibers of cervical nerve C1, conveyed by the hypoglossal nerve X11.

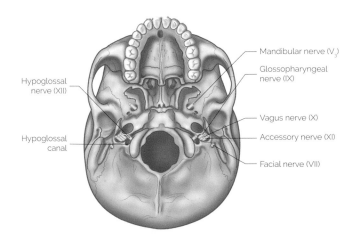

Hypoglossal nerve (XII)

Hypoglossal canal

Mandibular nerve (V$_3$)

Glossopharyngeal nerve (IX)

Vagus nerve (X)

Accessory nerve (XI)

Facial nerve (VII)

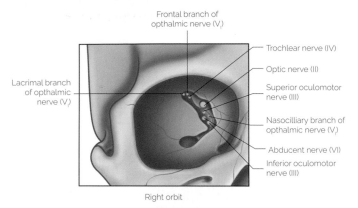

Frontal branch of opthalmic nerve (V₁)

Trochlear nerve (IV)

Optic nerve (II)

Superior oculomotor nerve (III)

Nasocilliary branch of opthalmic nerve (V₁)

Abducent nerve (VI)

Inferior oculomotor nerve (III)

Lacrimal branch of opthalmic nerve (V₁)

Right orbit

Cranial Nerves and Skull Passageways (External View)

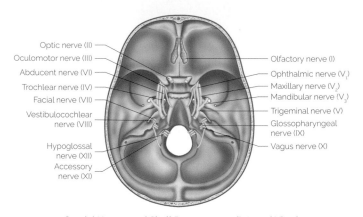

Optic nerve (II)

Oculomotor nerve (III)

Abducent nerve (VI)

Trochlear nerve (IV)

Facial nerve (VII)

Vestibulocochlear nerve (VIII)

Hypoglossal nerve (XII)

Accessory nerve (XI)

Olfactory nerve (I)

Ophthalmic nerve (V₁)

Maxillary nerve (V₂)

Mandibular nerve (V₃)

Trigeminal nerve (V)

Glossopharyngeal nerve (IX)

Vagus nerve (X)

Cranial Nerves and Skull Passageways (Internal View)

Cranial Nerve V—Trigeminal Nerve

Sensory Distribution

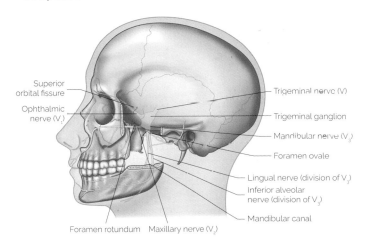

Superior orbital fissure

Ophthalmic nerve (V_1)

Trigeminal nerve (V)

Trigeminal ganglion

Mandibular nerve (V_3)

Foramen ovale

Lingual nerve (division of V_3)

Inferior alveolar nerve (division of V_3)

Mandibular canal

Foramen rotundum   Maxillary nerve (V_2)

Motor Distribution

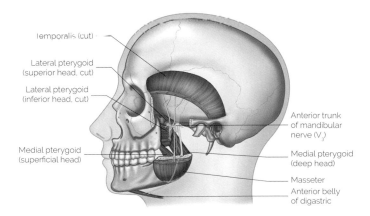

Temporalis (cut)

Lateral pterygoid (superior head, cut)

Lateral pterygoid (inferior head, cut)

Medial pterygoid (superficial head)

Anterior trunk of mandibular nerve (V_3)

Medial pterygoid (deep head)

Masseter

Anterior belly of digastric

Cranial Nerve VII—Facial Nerve

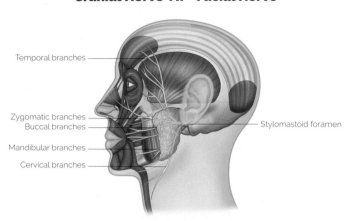

Temporal branches

Zygomatic branches
Buccal branches

Mandibular branches

Cervical branches

Stylomastoid foramen

Cranial Nerve XI—Accessory Nerve

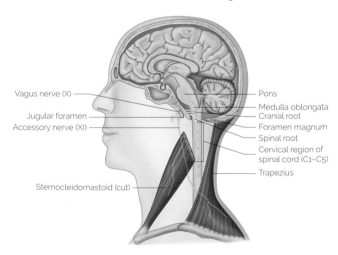

Vagus nerve (X)

Jugular foramen
Accessory nerve (XI)

Pons
Medulla oblongata
Cranial root
Foramen magnum
Spinal root
Cervical region of
spinal cord (C1–C5)

Sternocleidomastoid (cut)

Trapezius

Cervical Plexus

The cervical plexus is a network of nerves, formed
by the ventral rami of the four upper cervical nerves
(C1–4). The cervical plexus is located in the neck, deep to
sternocleidomastoid, and has two types of branch: cutaneous
and muscular. The **muscular branch** comprises: the **ansa
cervicalis nerve**, which innervates sternohyoid, sternothyroid,
thyrohyoid, and omohyoid; the **phrenic nerve**, which
innervates the diaphragm; and **segmental nerves**, which
innervate the middle and anterior scalenes. Furthermore,
longus colli, longus capitis, rectus capitis lateralis, and rectus
capitis anterior are also supplied via the cervical plexus. The
medial brachial cutaneous nerve innervates the skin on the
medial brachial side of the arm.

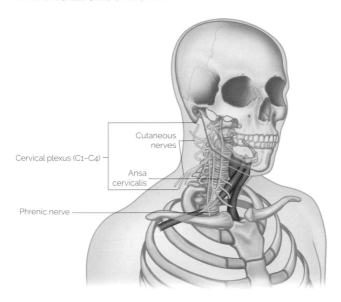

Cutaneous
nerves

Cervical plexus (C1–C4)

Ansa
cervicalis

Phrenic nerve

Brachial Plexus and Axillary Nerve

The **brachial plexus** is a network of nerves, formed by the anterior rami of the four lower cervical nerves (C5–8) and first thoracic nerve (T1). The brachial plexus is divided into roots (anterior rami of C5–8 and T1), trunks (superior, middle, inferior), divisions (each of the three trunks splitting in two, to create six divisions), cords (the six divisions regroup to form three cords—lateral, posterior, medial), and finally branches (nerves). Scalenus posterior, rhomboids, latissimus dorsi, supraspinatus, infraspinatus, subscapularis, teres major, and levator scapulae are innervated by the brachial plexus. The five main nerves originating from the brachial plexus are the axillary, median, musculocutaneous, ulnar, and radial nerves.

Axillary Nerve

The **axillary nerve** carries nerve fibers from C5 and C6, innervating the deltoid and teres minor.

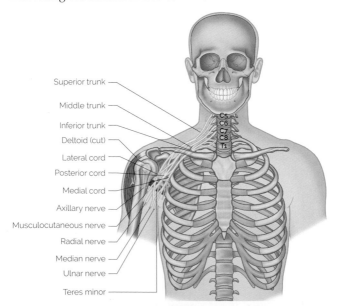

Superior trunk
Middle trunk
Inferior trunk
Deltoid (cut)
Lateral cord
Posterior cord
Medial cord
Axillary nerve
Musculocutaneous nerve
Radial nerve
Median nerve
Ulnar nerve
Teres minor

C5
C6
C7
C8
T1

Musculocutaneous Nerve

The fibers of the **musculocutaneous nerve** are derived from C5–7. It innervates coracobrachialis, biceps brachii, and brachialis. If the musculocutaneous nerve is damaged, the patient may present with weak flexion and supination of the forearm.

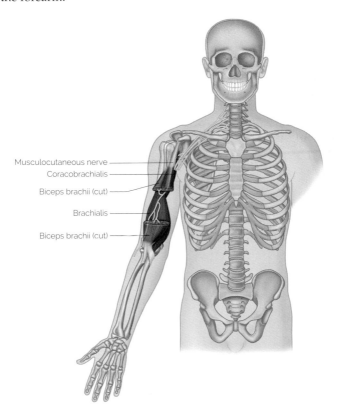

Musculocutaneous nerve

Coracobrachialis

Biceps brachii (cut)

Brachialis

Biceps brachii (cut)

Median Nerve

This nerve is derived from the anterior primary rami of C6, C7, C8, and T1. It gives off no branches in the arm but innervates all of the flexors in the forearm, except flexor carpi ulnaris and the medial half of flexor digitorum profundus (both supplied by the ulnar nerve). These forearm muscles are pronator teres, flexor carpi radialis, palmaris longus, flexor digitorum superficialis, flexor digitorum profundus (lateral half), flexor pollicis longus, and pronator quadratus. In the hand, the median nerve also innervates flexor pollicis brevis (superficial head), opponens pollicis, abductor pollicis brevis, and the first and second lumbricals.

The median nerve supplies sensation to the lateral palm, palmar skin, and the dorsal nail beds of the lateral three and a half digits. Compression of this nerve in the wrist, as it passes through the carpal tunnel, can cause carpal tunnel syndrome.

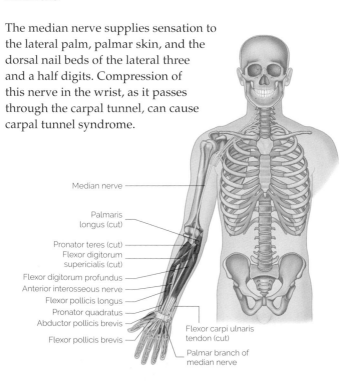

Median nerve

Palmaris longus (cut)

Pronator teres (cut)
Flexor digitorum supericialis (cut)

Flexor digitorum profundus
Anterior interosseous nerve
Flexor pollicis longus
Pronator quadratus
Abductor pollicis brevis

Flexor pollicis brevis

Flexor carpi ulnaris tendon (cut)

Palmar branch of median nerve

Ulnar Nerve

The nerve fibers of the **ulnar nerve** derive from C8 and T1. The nerve passes down through the arm and then winds under the medial epicondyle to enter the forearm and supply flexor carpi ulnaris and half of flexor digitorum profundus (the other half being supplied by the median nerve). In the lower forearm, the dorsal and palmar cutaneous branches are given off. The ulnar nerve then passes superficial to the flexor retinaculum after which it divides into terminal branches. The *superficial branch* ends as digital nerves supplying the skin of the little finger and the medial half of the ring finger. *The deep branch supplies* the hypothenar muscles, two lumbricals, the interossei and the adductor pollicis.

The ulnar nerve is the longest unprotected nerve in the human body and is therefore prone to injury. This tends to occur at the elbow (e.g. fracture of the medial epicondyle) or at the wrist from a laceration.

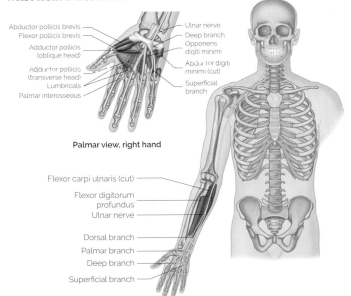

Abductor pollicis brevis
Flexor pollicis brevis
Adductor pollicis (oblique head)
Adductor pollicis (transverse head)
Lumbricals
Palmar interosseous

Ulnar nerve
Deep branch
Opponens digiti minimi
Abductor digiti minimi (cut)
Superficial branch

Palmar view, right hand

Flexor carpi ulnaris (cut)
Flexor digitorum profundus
Ulnar nerve
Dorsal branch
Palmar branch
Deep branch
Superficial branch

Radial Nerve

The fibers of the **radial nerve** are derived from C5–T1; the nerve subdivides into muscular and deep branches. The muscular branch innervates triceps brachii, anconeus, brachioradialis, and extensor carpi radialis longus. The deep branch innervates extensor carpi radialis brevis and supinator. The posterior interosseous nerve (a continuation of the deep branch) innervates extensor digitorum, extensor digiti minimi, extensor carpi ulnaris, abductor pollicis longus, extensor pollicis brevis, extensor pollicis longus, and extensor indicis.

The radial nerve sits in the spiral groove of the humerus and therefore a humeral shaft fracture may result in this nerve being damaged, leading to wrist-drop and loss of sensation of the skin over the anatomical snuffbox.

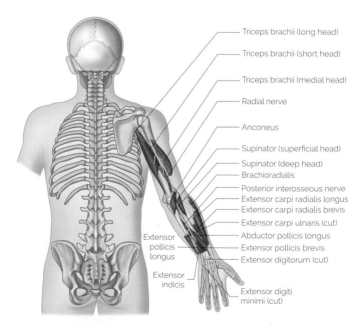

Triceps brachii (long head)
Triceps brachii (short head)
Triceps brachii (medial head)
Radial nerve
Anconeus
Supinator (superficial head)
Supinator (deep head)
Brachioradialis
Posterior interosseous nerve
Extensor carpi radialis longus
Extensor carpi radialis brevis
Extensor carpi ulnaris (cut)
Abductor pollicis longus
Extensor pollicis brevis
Extensor digitorum (cut)
Extensor pollicis longus
Extensor indicis
Extensor digiti minimi (cut)

Lumbar Plexus

The **lumbar plexus** forms part of the lumbosacral plexus, and is formed by the divisions of the first four lumbar nerves (L1–4) and the subcostal nerve (T12). Branches include: the **ilioinguinal** and **iliohypogastric nerves**, which innervate internal oblique and transversus abdominis; the **genitofemoral nerve**, which innervates cremaster; the **inferior gluteal nerve**, which innervates gluteus maximus; and the **superior gluteal nerve**, which innervates tensor fasciae latae, gluteus medius, and gluteus minimus, Also supplied via the lumbosacral plexus are piriformis (nerve to piriformis L5, S1), obturator internus (nerve to obturator L5, S1, 2), gemellus superior and inferior (nerve to obturator L5, S1, 2), and quadratus femoris (nerve to quadratus femoris L4–5). See also the obturator, femoral, sciatic, tibial, and common fibular nerves discussed below.

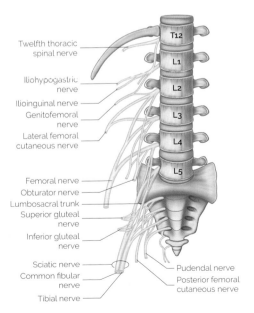

Twelfth thoracic spinal nerve

Iliohypogastric nerve

Ilioinguinal nerve

Genitofemoral nerve

Lateral femoral cutaneous nerve

Femoral nerve

Obturator nerve

Lumbosacral trunk

Superior gluteal nerve

Inferior gluteal nerve

Sciatic nerve

Common fibular nerve

Tibial nerve

T12

L1

L2

L3

L4

L5

Pudendal nerve

Posterior femoral cutaneous nerve

Sacral Plexus

The **sacral plexus** is a branching network of nerves that provides motor and sensory nerves to part of the pelvis, posterior thigh, most of the lower leg, and the entire foot. The sacral plexus is itself derived from the anterior rami of spinal nerves L4, L5, S1, S2, S3, and S4. Each of these anterior rami gives rise to anterior and posterior branches. The anterior branches supply flexor muscles of the lower limb, and posterior branches supply the extensor and abductor muscles.

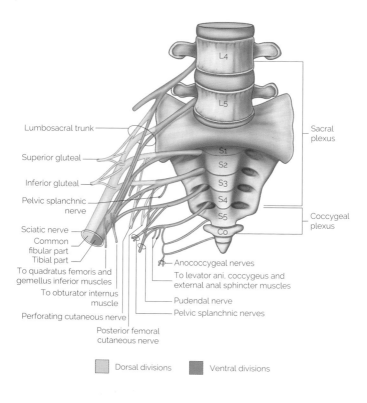

Lumbosacral trunk

Superior gluteal

Inferior gluteal

Pelvic splanchnic nerve

Sciatic nerve

Common fibular part

Tibial part

To quadratus femoris and gemellus inferior muscles

To obturator internus muscle

Perforating cutaneous nerve

Posterior femoral cutaneous nerve

L4

L5

S1

S2

S3

S4

S5

Co

Sacral plexus

Coccygeal plexus

Anococcygeal nerves

To levator ani, coccygeus and external anal sphincter muscles

Pudendal nerve

Pelvic splanchnic nerves

Dorsal divisions   Ventral divisions

Obturator Nerve

The **obturator nerve** originates from the ventral divisions of the second, third, and fourth lumbar nerves in the lumbar plexus and innervates obturator externus, adductor brevis, adductor magnus, adductor longus, gracilis, and pectineus (occasionally). Despite its name, the obturator nerve is not responsible for the innervation of obturator internus, which is supplied by the nerve to obturator internus from the sciatic nerve.

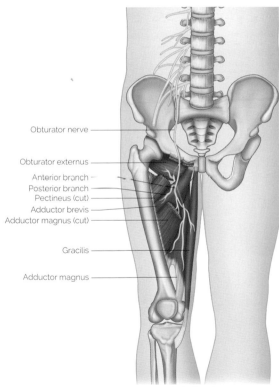

Obturator nerve

Obturator externus

Anterior branch
Posterior branch
Pectineus (cut)
Adductor brevis
Adductor magnus (cut)

Gracilis

Adductor magnus

Anterior view

Femoral Nerve

The **femoral nerve**, the largest branch of the lumbosacral plexus, is located in the thigh and not in the leg as some texts claim. It originates from the dorsal divisions of the ventral rami of the second, third, and fourth lumbar nerves (L2–4). In the femoral region, the nerve subdivides into the anterior and posterior divisions, before subdividing further into many smaller branches throughout the anterior and medial thigh. The **anterior division** innervates iliacus, sartorius, and pectineus, while the **posterior division** innervates rectus femoris, vastus lateralis, vastus medialis, and vastus intermedius.

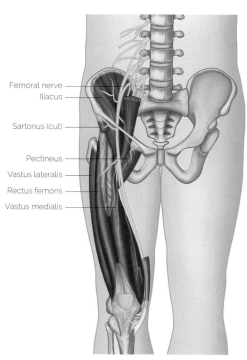

Femoral nerve
Iliacus

Sartorius (cut)

Pectineus
Vastus lateralis
Rectus femoris
Vastus medialis

Anterior view

Sciatic Nerve

The sciatic nerve is the longest and widest nerve in the human body. It is formed in the upper sacral plexus from the anterior primary rami of L4, L5, S1, S2, and S3. It passes out of the greater sciatic foramen, passing below piriformis. The sciatic nerve innervates biceps femoris, semimembranosus, and semitendinosus. True sciatic nerve damage can result in

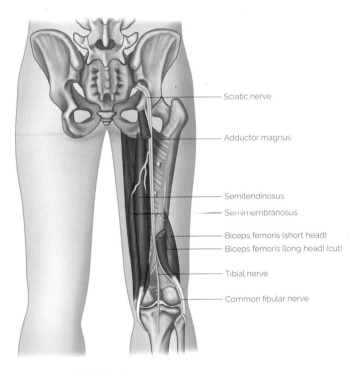

Sciatic nerve

Adductor magnus

Semitendinosus

Semimembranosus

Biceps femoris (short head)

Biceps femoris (long head) (cut)

Tibial nerve

Common fibular nerve

Posterior view

altered sensation, numbness, weakness, and pain. Depending on the source and level of irritation, the pain can be mild to severe. Sciatic nerve irritation usually occurs at the L5 or S1 level of the spine and only on one side. Pain can travel all the way to the foot and can affect normal motion, but with normal healing, the referred pain should dissipate and become more central. Unresolved chronic pain, especially of unknown origin, should be brought to the attention of the doctor or primary healthcare team.

At approximately mid thigh, the sciatic nerve divides into the tibial nerve and the common fibular nerve.

Tibial Nerve

The **tibial nerve** is a branch of the sciatic nerve, and innervates the muscles of the posterior compartment of the leg, including gastrocnemius, plantaris, soleus, flexor digitorum longus, tibialis posterior, popliteus, and flexor hallucis longus. One of its branches, the **medial plantar nerve**, innervates abductor hallucis, flexor digitorum brevis, flexor hallucis brevis, and the first lumbrical. The other branch, the **lateral plantar nerve**, innervates abductor digiti minimi, quadratus plantae, adductor hallucis, flexor digiti minimi brevis, plantar interossei, dorsal interossei, and the three lateral lumbricals.

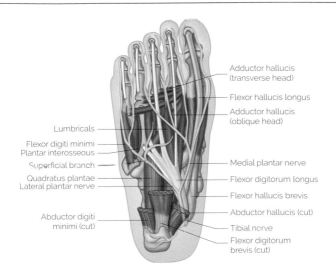

Adductor hallucis
(transverse head)

Flexor hallucis longus

Adductor hallucis
(oblique head)

Lumbricals

Flexor digiti minimi
Plantar interosseous

Superficial branch

Quadratus plantae
Lateral plantar nerve

Medial plantar nerve

Flexor digitorum longus

Flexor hallucis brevis

Abductor digiti
minimi (cut)

Abductor hallucis (cut)

Tibial nerve

Flexor digitorum
brevis (cut)

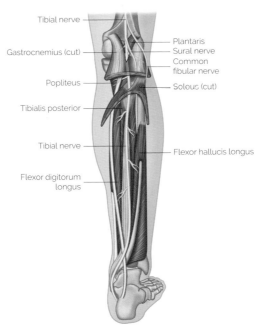

Tibial nerve

Gastrocnemius (cut)

Popliteus

Tibialis posterior

Tibial nerve

Flexor digitorum
longus

Plantaris
Sural nerve
Common
fibular nerve

Soleus (cut)

Flexor hallucis longus

Common Fibular Nerve

The **common fibular nerve** originates, via the sciatic nerve, from the dorsal branches of the fourth and fifth lumbar nerves (L4–5) and the first and second sacral nerves (S1–2). It divides into the superficial fibular nerve and the deep fibular nerve. The **superficial fibular nerve** innervates fibularis longus and fibularis brevis. The **deep fibular nerve** innervates tibialis anterior, extensor digitorum longus, fibularis tertius, extensor hallucis longus, extensor hallucis brevis, and extensor digitorum brevis.

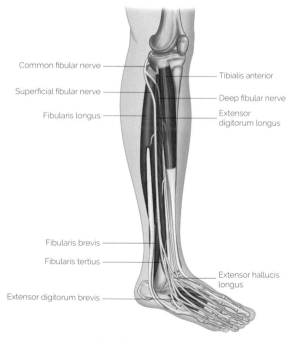

Common fibular nerve

Superficial fibular nerve

Fibularis longus

Tibialis anterior

Deep fibular nerve

Extensor digitorum longus

Fibularis brevis

Fibularis tertius

Extensor digitorum brevis

Extensor hallucis longus

Anterolateral view, right leg

Muscles Involved in Movement

Muscles of the Mandible

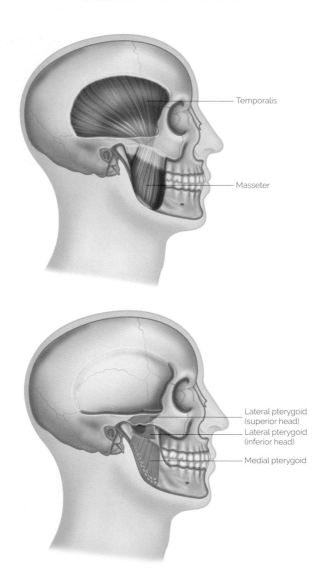

Temporalis

Masseter

Lateral pterygoid (superior head)

Lateral pterygoid (inferior head)

Medial pterygoid

➲ **Masseter**

Origin	Insertion	Nerve
Zygomatic arch and maxillary process of zygomatic bone.	Lateral surface of ramus of mandible.	Trigeminal V nerve (mandibular division).

➲ **Temporalis**

Origin	Insertion	Nerve
Bone of temporal fossa. Temporal fascia.	Coronoid process of mandible. Anterior border of ramus of mandible.	Deep temporal nerves from the trigeminal V nerve (mandibular division).

➲ **Lateral Pterygoid**

Origin	Insertion	Nerve
Superior head: Roof of infratemporal fossa. Inferior head: Lateral surface of lateral pterygoid plate of pterygoid process.	Superior head: Capsule and articular disc of temporomandibular joint. Inferior head: Neck of mandible.	Trigeminal V nerve (mandibular division).

➲ **Medial Pterygoid**

Origin	Insertion	Nerve
Deep head: Medial surface of lateral pterygoid plate of pterygoid process. Superficial head: Pyramidal process of palatine bone. Tuberosity of maxilla.	Medial surface of ramus and angle of mandible.	Trigeminal V nerve (mandibular division).

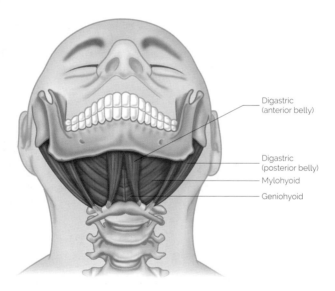

Digastric
(anterior belly)

Digastric
(posterior belly)

Mylohyoid

Geniohyoid

⊃ **Mylohyoid**

Origin	Insertion	Nerve
Mylohyoid line on inner surface of mandible.	Hyoid bone.	Mylohyoid nerve from inferior alveolar nerve, a branch of trigeminal V nerve (mandibular division).

⊃ **Geniohyoid**

Origin	Insertion	Nerve
Inferior mental spines of mandible.	Hyoid bone.	C1.

⊃ **Digastric**

Origin	Insertion	Nerve
Anterior belly: digastric fossa on lower border of mandible. Posterior belly: mastoid notch of temporal bone.	Body of hyoid bone via a fascial sling over an intermediate tendon.	Anterior belly: mylohyoid nerve, from trigeminal V nerve (mandibular division). Posterior belly: facial VII nerve (digastric branch).

Movements of the Mandible

Elevation	Depression	Protraction	Retraction	Chewing
Temporalis (anterior fibers), masseter, medial pterygoid	Lateral pterygoid, digastric, mylohyoid, geniohyoid	Lateral pterygoid, medial pterygoid, masseter (superficial fibers)	Temporalis (horizontal fibers), digastric	Lateral pterygoid, medial pterygoid, masseter, temporalis

Muscles of the Neck

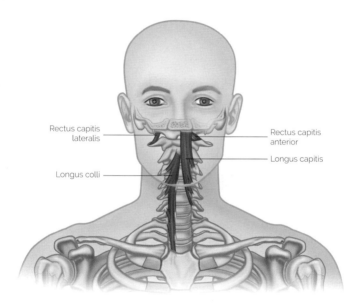

Rectus capitis
lateralis

Rectus capitis
anterior

Longus capitis

Longus colli

⊃ **Longus Colli**

Origin	Insertion	Nerve
Superior oblique: transverse processes of third to fifth cervical vertebrae (C3–5).

Inferior oblique: anterior surface of first two or three thoracic vertebral bodies.

Vertical: anterior surface of upper three thoracic and lower three cervical vertebral bodies. | Superior oblique: anterior arch of atlas.

Inferior oblique: transverse processes of fifth and sixth cervical vertebrae (C5–6).

Vertical: surface of bodies of second to fourth cervical vertebrae (C2–4). | Ventral rami of cervical nerves C2–6. |

⊃ **Longus Capitis**

Origin	Insertion	Nerve
Transverse processes of third to sixth cervical vertebrae (C3–6).	Inferior surface of basilar part of occipital bone.	Ventral rami of cervical nerves C1–3.

⊃ **Rectus Capitis Anterior**

Origin	Insertion	Nerve
Anterior surface of lateral mass of atlas.	Basilar part of occipital bone, anterior to occipital condyle.	Loop between ventral rami of cervical nerves C1, 2.

⊃ **Rectus Capitis Lateralis**

Origin	Insertion	Nerve
Transverse process of atlas.	Jugular process of occipital bone.	Loop between ventral rami of cervical nerves C1, 2.

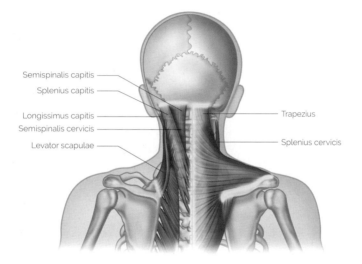

Semispinalis capitis

Splenius capitis

Longissimus capitis

Semispinalis cervicis

Levator scapulae

Trapezius

Splenius cervicis

➲ Longissimus

Origin	Insertion	Nerve
Cervicis: transverse processes of upper four or five thoracic vertebrae (T1–5). Capitis: transverse processes of upper five thoracic vertebrae (T1–5). Articular processes of lower three cervical vertebrae (C5–7).	Cervicis: transverse processes of second to sixth cervical vertebrae (C2–6). Capitis: posterior part of mastoid process of temporal bone.	Dorsal rami of spinal nerves C1–S1.

➲ Splenius Capitis and Splenius Cervicis

Origin	Insertion	Nerve
Capitis: lower part of ligamentum nuchae. Spinous processes of seventh cervical vertebra (C7) and upper three or four thoracic vertebrae (T1–4). Cervicis: spinous processes of third to sixth thoracic vertebrae (T3–6).	Capitis: posterior aspect of mastoid process of temporal bone. Lateral part of superior nuchal line, deep to attachment of sternocleidomastoid. Cervicis: posterior tubercles of transverse processes of upper two or three cervical vertebrae (C1–3).	Capitis: posterior rami of middle cervical nerves. Cervicis: posterior rami of lower cervical nerves.

➲ Semispinalis

Origin	Insertion	Nerve
Cervicis: transverse processes of upper five or six thoracic vertebrae (T1–6). Capitis: transverse processes of lower four cervical and upper six or seven thoracic vertebrae (C4–T7).	Cervicis: spinous processes of second to fifth cervical vertebrae (C2–5). Capitis: between superior and inferior nuchal lines of occipital bone.	Dorsal rami of thoracic and cervical spinal nerves.

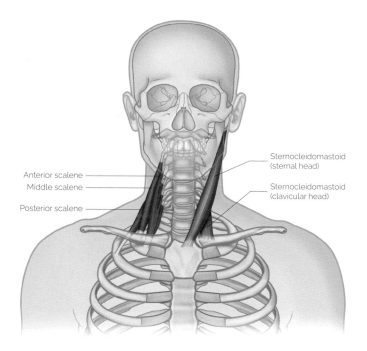

Anterior scalene

Middle scalene

Posterior scalene

Sternocleidomastoid
(sternal head)

Sternocleidomastoid
(clavicular head)

➲ **Scalenes**

Origin	Insertion	Nerve
Anterior: transverse processes of third to sixth cervical vertebrae (C3–6). Middle: posterior tubercles of transverse processes of lower six cervical vertebrae (C2–7). Posterior: posterior tubercles of transverse processes of fourth to sixth cervical vertebrae (C4–6).	Anterior: scalene tubercle on inner border of first rib. Middle: upper surface of first rib, behind groove for subclavian artery. Posterior: outer surface of second rib.	Anterior: ventral rami of cervical nerves C4–7. Middle: ventral rami of cervical nerves C3–7. Posterior: ventral rami of cervical nerves C5–7.

➲ **Sternocleidomastoid**

Origin	Insertion	Nerve
Sternal head: anterior surface of manubrium of sternum. Clavicular head: upper surface of medial third of clavicle.	Sternal head: lateral third of superior nuchal line of occipital bone. Clavicular head: outer surface of mastoid process of temporal bone.	Accessory XI nerve, with sensory supply for proprioception from cervical nerves C2 and C3.

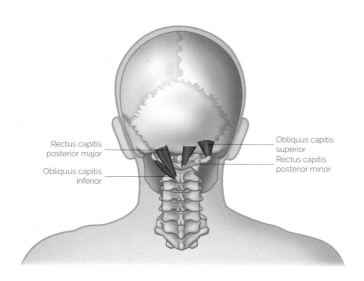

Rectus capitis posterior major

Obliquus capitis inferior

Obliquus capitis superior

Rectus capitis posterior minor

⮑ Rectus Capitis Posterior Major

Origin	Insertion	Nerve
Spinous process of axis.	Lateral portion of occipital bone below inferior nuchal line.	Suboccipital nerve (dorsal ramus of first cervical nerve C1).

⮑ Rectus Capitis Posterior Minor

Origin	Insertion	Nerve
Posterior tubercle of atlas.	Medial portion of occipital bone below inferior nuchal line.	Suboccipital nerve (dorsal ramus of first cervical nerve C1).

⇒ **Obliquus Capitis Inferior**

Origin	Insertion	Nerve
Spinous process of axis.	Transverse process of atlas.	Suboccipital nerve (dorsal ramus of first cervical nerve C1).

⇒ **Obliquus Capitis Superior**

Origin	Insertion	Nerve
Transverse process of atlas.	Area between inferior and superior nuchal lines on occipital bone.	Suboccipital nerve (dorsal ramus of first cervical nerve C1).

Movements of the Neck

⇒ **Atlanto-Occipital and Atlanto-Axial Joints**

Flexion	Extension	Rotation and Lateral Flexion
Longus capitis, rectus capitis anterior, sternocleidomastoid (anterior fibers)	Semispinalis capitis, splenius capitis, rectus capitis posterior major, rectus capitis posterior minor, obliquus capitis superior, longissimus capitis, trapezius, sternocleidomastoid (posterior fibers)	Sternocleidomastoid, obliquus capitis inferior, obliquus capitis superior, rectus capitis lateralis, longissimus capitis, splenius capitis

⇒ **Intervertebral Joints (Cervical Region)**

Flexion	Extension	Rotation and Lateral Flexion
Longus colli, longus capitis, sternocleido-mastoid	Longissimus cervicis, longissimus capitis, splenius capitis, splenius cervicis, semispinalis cervicis, semispinalis capitis, trapezius, interspinales, iliocostalis cervicis	Longissimus cervicis, longissimus capitis, splenius capitis, splenius cervicis, multifidus, longus colli, scalenus anterior, scalenus medius, scalenus posterior, sternocleidomastoid, levator scapulae, iliocostalis cervicis, intertransversarii

Muscles of the Trunk

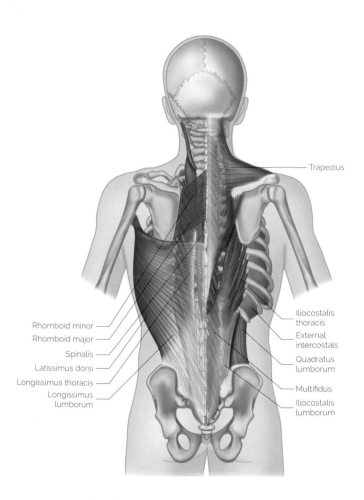

Trapezius

Rhomboid minor
Rhomboid major
Spinalis
Latissimus dorsi
Longissimus thoracis
Longissimus lumborum

Iliocostalis thoracis
External intercostals
Quadratus lumborum
Multifidus
Iliocostalis lumborum

⊃ Trapezius

Origin	Insertion	Nerve
Medial third of superior nuchal line of occipital bone. External occipital protuberance. Ligamentum nuchae. Spinous processes and supraspinous ligaments of seventh cervical vertebra (C7) and all thoracic vertebrae (T1–12).	Posterior border of lateral third of clavicle. Medial border of acromion. Upper border of crest of spine of scapula, and tubercle on this crest.	Motor supply: accessory XI nerve. Sensory supply (proprioception): ventral ramus of cervical nerves C2, 3, 4.

⊃ Iliocostalis

Origin	Insertion	Nerve
Lumborum: lateral and medial sacral crests. Medial part of iliac crests. Thoracis: angles of lower six ribs, medial to iliocostalis lumborum.	Lumborum: angles of lower six or seven ribs. Thoracis: angles of upper six ribs and transverse process of seventh cervical vertebra (C7).	Dorsal rami of cervical, thoracic and lumbar spinal nerves.

⊃ Multifidus

Origin	Insertion	Nerve
Sacrum, origin of erector spinae, PSIS, mammillary processes of lumbar vertebrae, transverse processes of thoracic vertebrae, articular processes of C3–C7.	Base of spinous processes of all vertebrae from L5 to C2.	Dorsal rami of spinal nerves.

⊃ Rotatores

Origin	Insertion	Nerve
Transverse process of each vertebra.	Base of spinous process of adjoining vertebra above.	Dorsal rami of spinal nerves.

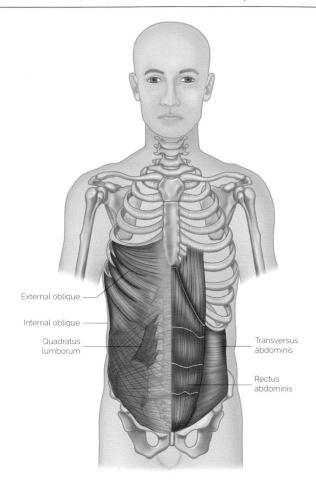

External oblique

Internal oblique

Quadratus lumborum

Transversus abdominis

Rectus abdominis

◑ External Oblique

Origin	Insertion	Nerve
Muscular slips from the outer surfaces of the lower eight ribs (ribs V to XII).	Lateral lip of iliac crest; aponeurosis ending in linea alba.	Ventral rami of thoracic nerves T7–12.

⊃ **Internal Oblique**

Origin	Insertion	Nerve
Iliac crest. Lateral two-thirds of inguinal ligament. Thoracolumbar fascia.	Inferior borders of bottom three or four ribs. Linea alba via an abdominal aponeurosis. Pubic crest and pectineal line.	Ventral rami of thoracic nerves T7–12, and L1.

⊃ **Transversus Abdominis**

Origin	Insertion	Nerve
Anterior two-thirds of iliac crest. Lateral third of inguinal ligament. Thoracolumbar fascia. Costal cartilages of lower six ribs (ribs VII to XII).	Aponeurosis ending in linea alba; pubic crest and pectineal line.	Ventral rami of thoracic nerves T7–12, and L1.

⊃ **Rectus Abdominis**

Origin	Insertion	Nerve
Pubic crest, pubic tubercle, and pubic symphysis.	Anterior surface of xiphoid process. Fifth, sixth, and seventh costal cartilages.	Ventral rami of thoracic nerves T7–12.

⊃ **Quadratus Lumborum**

Origin	Insertion	Nerve
Posterior part of iliac crest. Iliolumbar ligament.	Medial part of lower border of twelfth rib. Transverse processes of upper four lumbar vertebrae (L1–4).	Ventral rami of T12 and L1–4.

Movements of the Intervertebral Joints (Thoracic Region)

Flexion	Extension	Rotation and Lateral Flexion
Muscles of anterior abdominal wall	Erector spinae, quadratus lumborum, trapezius	Iliocostalis lumborum, iliocostalis thoracis, multifidus, rotatores, intertransversarii, quadratus lumborum, psoas major, muscles of anterior abdominal wall

Muscles of the Shoulder

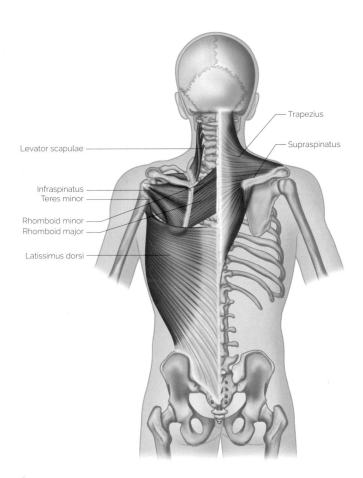

Trapezius

Supraspinatus

Levator scapulae

Infraspinatus

Teres minor

Rhomboid minor

Rhomboid major

Latissimus dorsi

⊃ Levator Scapulae

Origin	Insertion	Nerve
Posterior tubercles of transverse processes of first three or four cervical vertebrae (C1–4).	Medial (vertebral) border of scapula, between superior angle and spine of scapula.	Dorsal scapular nerve C4, 5, and cervical nerves C3, 4.

⊃ Rhomboids

Origin	Insertion	Nerve
Minor: spinous processes of seventh cervical and first thoracic vertebrae. Lower part of ligamentum nuchae. Major: spinous processes of second to fifth thoracic vertebrae (T2–5).	Minor: medial border of scapula at the level of spine of scapula. Major: medial border of scapula, between spine of scapula and inferior angle.	Dorsal scapular nerve C4, 5.

⊃ Serratus Anterior

Origin	Insertion	Nerve
Outer surfaces and superior borders of upper eight or nine ribs, and fascia covering their intercostal spaces.	Anterior (costal) surface of medial border of scapula and inferior angle of scapula.	Long thoracic nerve C5–7.

⊃ Latissimus Dorsi

Origin	Insertion	Nerve
Thoracolumbar fascia, which is attached to spinous processes of lower six thoracic vertebrae and all lumbar and sacral vertebrae (T7–S5) and to intervening supraspinous ligaments. Posterior part of iliac crest. Lower three or four ribs. Inferior angle of scapula.	Floor of intertubercular sulcus (bicipital groove) of humerus.	Thoracodorsal nerve C6, 7, 8, from the posterior cord of brachial plexus.

⊃ Supraspinatus

Origin	Insertion	Nerve
Supraspinous fossa of scapula.	Upper aspect of greater tubercle of humerus. Capsule of shoulder joint.	Suprascapular nerve C5 and 6, from upper trunk of brachial plexus.

⊃ Infraspinatus

Origin	Insertion	Nerve
Infraspinous fossa of scapula.	Middle facet on greater tubercle of humerus. Capsule of shoulder joint.	Suprascapular nerve C5 and 6, from upper trunk of brachial plexus.

⊃ Teres Minor

Origin	Insertion	Nerve
Upper two-thirds of lateral border of dorsal surface of scapula.	Lower facet on greater tubercle of humerus. Capsule of shoulder joint.	Axillary nerve C5, 6, from posterior cord of brachial plexus.

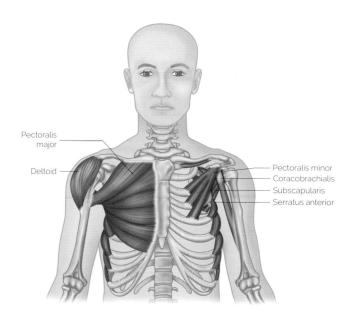

Pectoralis major

Deltoid

Pectoralis minor
Coracobrachialis
Subscapularis
Serratus anterior

⊃ Pectoralis Minor

Origin	Insertion	Nerve
Outer surfaces of third, fourth, and fifth ribs, and fascia of the corresponding intercostal spaces.	Coracoid process of scapula.	Medial pectoral nerve, with fibers from a communicating branch of lateral pectoral nerve, C5–8, T1.

➲ Pectoralis Major

Origin	Insertion	Nerve
Clavicular head: medial half or two-thirds of front of clavicle. Sternocostal portion: anterior surface of sternum. Upper seven costal cartilages. Aponeurosis of external oblique. Sternal end of sixth rib.	Lateral lip of intertubercular sulcus (bicipital groove) of humerus.	Medial and lateral pectoral nerves. Clavicular head: C5, 6. Sternocostal head: C6–8, T1.

➲ Deltoid

Origin	Insertion	Nerve
Anterior fibers: anterior border and superior surface of lateral third of clavicle. Middle fibers: lateral border of acromion process. Posterior fibers: lower lip of crest of spine of scapula.	Deltoid tuberosity of humerus.	Axillary nerve C5, 6, from the posterior cord of brachial plexus.

➲ Subscapularis

Origin	Insertion	Nerve
Subscapular fossa and groove along lateral border of anterior surface of scapula.	Lesser tubercle of humerus. Capsule of shoulder joint.	Upper and lower subscapular nerves C5, 6, 7, from posterior cord of brachial plexus.

➲ Teres Major

Origin	Insertion	Nerve
Oval area on lower third of posterior surface of lateral border of scapula.	Medial lip of intertubercular sulcus (bicipital groove) of humerus.	Lower subscapular nerve C5, 6, 7, from posterior cord of brachial plexus.

➲ Coracobrachialis

Origin	Insertion	Nerve
Tip of coracoid process of scapula.	Medial aspect of humerus at mid-shaft.	Musculocutaneous nerve C5–7.

Movements of the Shoulder Girdle

Elevation	Depression	Protraction	Retraction	Lateral Displacement of Inferior Angle of Scapula	Medial Displacement of Inferior Angle of Scapula
Trapezius (upper fibers), levator scapulae, rhomboid minor, rhomboid major, sternocleidomastoid	Trapezius (lower fibers), pectoralis minor, pectoralis major (sternocostal portion), latissimus dorsi	Serratus anterior, pectoralis minor, pectoralis major	Trapezius (middle fibers), rhomboid minor, rhomboid major, latissimus dorsi	Serratus anterior, trapezius (upper and lower fibers)	Pectoralis minor, rhomboid minor, rhomboid major, latissimus dorsi

Movements of the Shoulder Joint

Flexion	Extension	Abduction	Adduction	Lateral Rotation	Medial Rotation	Horizontal Flexion	Horizontal Extension
Anterior deltoid, pectoralis major (clavicular portion; sternocostal portion flexes the extended humerus as far as the position of rest), biceps brachii, coracobrachialis	Posterior deltoid, teres major (of flexed humerus), latissimus dorsi (of flexed humerus), pectoralis major (sternocostal portion of flexed humerus), triceps brachii (long head to position of rest)	Middle deltoid, supraspinatus, biceps brachii (long head)	Pectoralis major, teres major, latissimus dorsi, triceps brachii (long head), coracobrachialis	Posterior deltoid, infraspinatus, teres minor	Pectoralis major, teres major, latissimus dorsi, anterior deltoid, subscapularis	Anterior deltoid, pectoralis major, subscapularis	Posterior deltoid, infraspinatus

Muscles of the Elbow and Radioulnar Joint

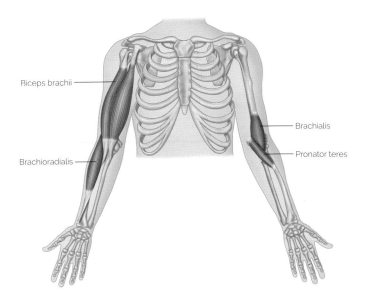

Biceps brachii

Brachialis

Brachioradialis

Pronator teres

➲ Biceps Brachii

Origin	Insertion	Nerve
Short head: tip of coracoid process of scapula. Long head: supraglenoid tubercle of scapula.	Tuberosity of radius.	Musculocutaneous nerve C5, 6.

➲ Brachialis

Origin	Insertion	Nerve
Lower (distal) two-thirds of anterior aspect of humerus.	Tuberosity of ulna.	Musculocutaneous nerve C5, 6.

➲ Pronator Teres

Origin	Insertion	Nerve
Humeral head: lower third of medial supracondylar ridge and common flexor origin on anterior aspect of medial epicondyle of humerus. Ulnar head: medial border of coronoid process of ulna.	Mid-lateral surface of radius (pronator tuberosity).	Median nerve C6, 7.

➲ Brachioradialis

Origin	Insertion	Nerve
Upper two-thirds of anterior aspect of lateral supracondylar ridge of humerus.	Lower lateral end of radius, just above styloid process.	Radial nerve C5, 6.

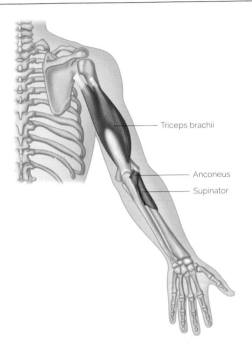

- Triceps brachii
- Anconeus
- Supinator

➲ Triceps Brachii

Origin	Insertion	Nerve
Long head: infraglenoid tubercle of scapula. Lateral head: upper half of posterior surface of shaft of humerus (above and lateral to radial groove). Medial head: lower half of posterior surface of shaft of humerus (below and medial to radial groove).	Posterior part of olecranon process of ulna.	Radial nerve C6–8.

⊃ **Anconeus**

Origin	Insertion	Nerve
Posterior part of lateral epicondyle of humerus.	Lateral surface of olecranon process and upper portion of posterior surface of ulna.	Radial nerve C6–8.

⊃ **Supinator**

Origin	Insertion	Nerve
Lateral epicondyle of humerus. Radial collateral and anular ligaments. Supinator crest of ulna.	Dorsal and lateral surfaces of upper third of radius.	Posterior interosseous nerve, a continuation of the deep branch of radial nerve C6, 7.

Movements of the Elbow Joint

Flexion	Extension
Brachialis, biceps brachii, brachioradialis, extensor carpi radialis longus, pronator teres, flexor carpi radialis	Triceps brachii, anconeus

Movements of the Radioulnar Joints

Supination	Pronation
Supinator, biceps brachii, extensor pollicis longus	Pronator quadratus, pronator teres, flexor carpi radialis

Muscles of the Radiocarpal and Midcarpal Joints

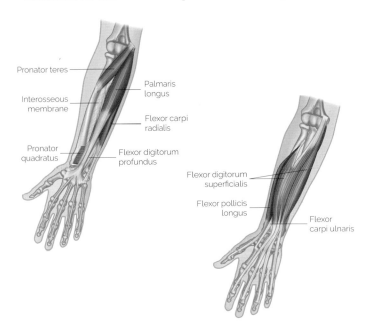

⮑ Flexor Carpi Ulnaris

Origin	Insertion	Nerve
Humeral head: common flexor origin on medial epicondyle of humerus. Ulnar head: medial border of olecranon. Posterior border of upper two-thirds of ulna.	Pisiform bone. Hook of hamate. Base of fifth metacarpal.	Ulnar nerve C7, 8, T1.

⮑ Palmaris Longus

Origin	Insertion	Nerve
Common flexor origin on anterior aspect of medial epicondyle of humerus.	Palmar aponeurosis of hand.	Median nerve C7, 8.

⊃ **Flexor Carpi Radialis**

Origin	Insertion	Nerve
Common flexor origin on anterior aspect of medial epicondyle of humerus.	Front of bases of second and third metacarpal bones.	Median nerve C6, 7.

⊃ **Flexor Digitorum Superficialis**

Origin	Insertion	Nerve
Humeroulnar head: long linear origin from common flexor tendon on medial epicondyle of humerus. Medial border of coronoid process of ulna. Radial head: upper two-thirds of anterior border of radius.	Four tendons each divide into two slips, each of which insert into the sides of the middle phalanges of the four fingers.	Median nerve C8, T1.

⊃ **Flexor Digitorum Profundus**

Origin	Insertion	Nerve
Upper two-thirds of medial and anterior surfaces of ulna, reaching up onto medial side of olecranon process. Interosseous membrane.	Anterior surface of base of distal phalanges.	Medial half of muscle, destined for the little and ring fingers: ulnar nerve C8, T1. Lateral half of muscle, destined for the index and middle fingers: anterior interosseous branch of median nerve C8, T1.

⊃ **Flexor Pollicis Longus**

Origin	Insertion	Nerve
Middle part of anterior surface of shaft of radius. Interosseous membrane.	Palmar surface of base of distal phalanx of thumb.	Anterior interosseous branch of median nerve C7, 8.

⊃ **Pronator Quadratus**

Origin	Insertion	Nerve
Distal quarter of anterior surface of shaft of ulna.	Lateral side of distal quarter of anterior surface of shaft of radius.	Anterior interosseous branch of median nerve C7, 8.

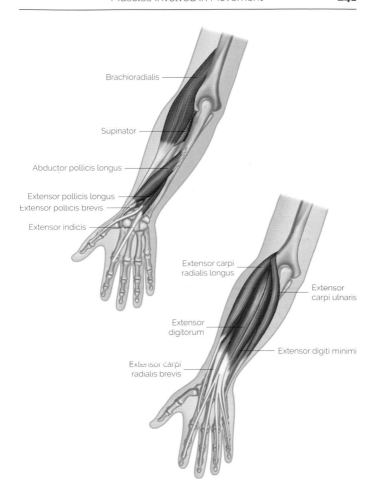

⊃ Extensor Carpi Radialis Longus

Origin	Insertion	Nerve
Lower (distal) third of lateral supracondylar ridge of humerus.	Dorsal surface of base of second metacarpal bone, on its radial side.	Radial nerve C6, 7.

⇨ **Extensor Carpi Radialis Brevis**

Origin	Insertion	Nerve
Common extensor tendon from lateral epicondyle of humerus.	Dorsal surface of second and third metacarpals.	Radial nerve C7, 8.

⇨ **Extensor Digitorum**

Origin	Insertion	Nerve
Common extensor tendon from lateral epicondyle of humerus.	Dorsal surfaces of all the phalanges of the four fingers.	Deep radial (posterior interosseous) nerve C7, 8.

⇨ **Extensor Digiti Minimi**

Origin	Insertion	Nerve
Common extensor tendon from lateral epicondyle of humerus.	Extensor hood of little finger.	Deep radial (posterior interosseous) nerve C7, 8.

⇨ **Extensor Carpi Ulnaris**

Origin	Insertion	Nerve
Common extensor tendon from lateral epicondyle of humerus. Aponeurosis from mid-posterior border of ulna.	Medial side of base of fifth metacarpal.	Deep radial (posterior interosseous) nerve C7, 8.

⇨ **Abductor Pollicis Longus**

Origin	Insertion	Nerve
Posterior surface of shaft of ulna, distal to origin of supinator. Interosseous membrane. Posterior surface of middle third of shaft of radius.	Lateral side of base of first metacarpal.	Deep radial (posterior interosseous) nerve C7, 8.

⊃ **Extensor Pollicis Brevis**

Origin	Insertion	Nerve
Posterior surface of radius, distal to origin of abductor pollicis longus. Adjacent part of interosseous membrane.	Base of dorsal surface of proximal phalanx of thumb.	Deep radial (posterior interosseous) nerve C7, 8.

⊃ **Extensor Pollicis Longus**

Origin	Insertion	Nerve
Middle third of posterior surface of ulna. Interosseous membrane.	Dorsal surface of base of distal phalanx of thumb.	Deep radial (posterior interosseous) nerve C7, 8.

⊃ **Extensor Indicis**

Origin	Insertion	Nerve
Posterior surface of ulna. Adjacent part of interosseous membrane.	Extensor hood of index finger.	Deep radial (posterior interosseous) nerve C7, 8.

Movements of the Radiocarpal and Midcarpal Joints

Flexion	Extension	Abduction	Adduction
Flexor carpi radialis, flexor carpi ulnaris, palmaris longus, flexor digitorum superficialis, flexor digitorum profundus, flexor pollicis longus, abductor pollicis longus, extensor pollicis brevis	Extensor carpi radialis brevis, extensor carpi radialis longus, extensor carpi ulnaris, extensor digitorum, extensor indicis, extensor pollicis longus, extensor digiti minimi	Extensor carpi radialis brevis, extensor carpi radialis longus, flexor carpi radialis, abductor pollicis longus, extensor pollicis longus, extensor pollicis brevis	Flexor carpi ulnaris, extensor carpi ulnaris

Movements of the Metacarpophalangeal Joints of the Fingers

Flexion	Extension	Abduction and Adduction	Rotation
Flexor digitorum profundus, flexor digitorum superficialis, lumbricales, interossei, flexor digiti minimi, abductor digiti minimi, palmaris longus (through palmar aponeurosis)	Extensor digitorum, extensor indicis, extensor digiti minimi	Interossei, abductor digiti minimi, lumbricales (may assist in radial deviation, extensor digitorum (abducts by hyperextending; tendon to index radially deviates), flexor digitorum profundus (adducts by flexing), flexor digitorum superficialis (adducts by flexing)	Lumbricales, interossei (movement is slight except index; only effective when phalanx is flexed), opponens digiti minimi (rotates little finger at carpometacarpal joint)

Movements of the Interphalangeal Joints of the Fingers

Flexion	Extension
Flexor digitorum profundus (both joints), flexor digitorum superficialis (proximal joint only)	Extensor digitorum, extensor digiti minimi, extensor indicis, lumbricales, interossei

Movements of the Interphalangeal Joint of the Thumb

Flexion	Extension
Flexor pollicis longus	Extensor pollicis longus

Muscles of the Hand

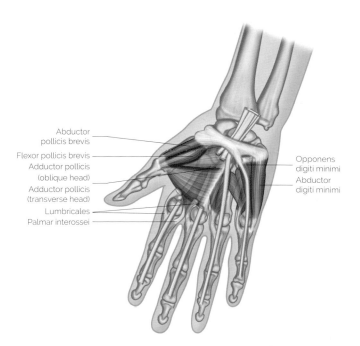

Abductor pollicis brevis
Flexor pollicis brevis
Adductor pollicis (oblique head)
Adductor pollicis (transverse head)
Lumbricales
Palmar interossei

Opponens digiti minimi
Abductor digiti minimi

⊃ Palmaris Brevis

Origin	Insertion	Nerve
Palmar aponeurosis. Flexor retinaculum.	Skin on ulnar border of hand.	Ulnar nerve C8, T1.

⊃ Flexor Pollicis Brevis

Origin	Insertion	Nerve
Tubercle of trapezium and flexor retinaculum.	Proximal phalanx of thumb.	Median nerve (C8, T1).

➲ Dorsal Interossei

Origin	Insertion	Nerve
By two heads, each from adjacent sides of metacarpals. Therefore, each of the dorsal interossei occupies an interspace between adjacent metacarpals.	Into extensor hood and to base of proximal phalanx as follows: First: lateral side of index finger, mainly to base of proximal phalanx. Second: lateral side of middle finger. Third: medial side of middle finger, mainly into extensor expansion. Fourth: medial side of ring finger.	Ulnar nerve C8, T1.

➲ Palmar Interossei

Origin	Insertion	Nerve
First: medial side of base of first metacarpal. Second: medial side of shaft of second metacarpal. Third: lateral side of shaft of fourth metacarpal. Fourth: lateral side of shaft of fifth metacarpal.	Primarily into extensor hood of respective digit, with possible attachment to base of proximal phalanx as follows: First: medial side of proximal phalanx of thumb. Second: medial side of proximal phalanx of index finger. Third: lateral side of proximal phalanx of ring finger. Fourth: lateral side of proximal phalanx of little finger.	Ulnar nerve C8, T1.

➲ Adductor Pollicis

Origin	Insertion	Nerve
Oblique fibers: anterior surfaces of second and third metacarpals and capitate. Transverse fibers: palmar surface of third metacarpal bone.	Medial side of base of proximal phalanx of thumb.	Deep ulnar nerve C8, T1.

⊃ Lumbricals

Origin	Insertion	Nerve
Tendons of flexor digitorum profundus in palm.	Extensor hoods of index, ring, middle, and little fingers.	Lateral lumbricals (first and second): median nerve. Medial lumbricals (third and fourth): ulnar nerve.

⊃ Abductor Digiti Minimi

Origin	Insertion	Nerve
Pisiform bone. Tendon of flexor carpi ulnaris.	Proximal phalanx of little finger.	Ulnar nerve C8, T1.

⊃ Opponens Digiti Minimi

Origin	Insertion	Nerve
Hook of hamate. Anterior surface of flexor retinaculum.	Medial border of fifth metacarpal.	Ulnar nerve C8, T1.

⊃ Flexor Digiti Minimi Brevis

Origin	Insertion	Nerve
Hook of hamate. Anterior surface of flexor retinaculum.	Proximal phalanx of little finger.	Ulnar nerve C8, T1.

⊃ Abductor Pollicis Brevis

Origin	Insertion	Nerve
Flexor retinaculum. Tubercle of trapezium. Tubercle of scaphoid.	Proximal phalanx and extensor hood of thumb.	Median nerve (C8, T1).

⊃ Opponens Pollicis

Origin	Insertion	Nerve
Flexor retinaculum. Tubercle of trapezium.	Entire length of radial border of first metacarpal.	Median nerve (C8, T1)

Movements of the Carpometacarpal Joint of the Thumb

Flexion	Extension	Abduction	Adduction	Opposition
Flexor pollicis brevis, flexor pollicis longus, opponens pollicis	Extensor pollicis brevis, extensor pollicis longus, abductor pollicis longus	Abductor pollicis brevis, abductor pollicis longus	Adductor pollicis, dorsal interossei (first only), extensor pollicis longus (in full extension/abduction), flexor pollicis longus (in full extension/abduction)	Opponens pollicis, abductor pollicis brevis, flexor pollicis brevis, flexor pollicis longus, adductor pollicis

Movements of the Metacarpophalangeal Joint of the Thumb

Flexion	Extension	Abduction	Adduction
Flexor pollicis brevis, flexor pollicis longus, abductor pollicis brevis	Extensor pollicis brevis, extensor pollicis longus	Abductor pollicis brevis	Adductor pollicis

Muscles of the Hip Joint

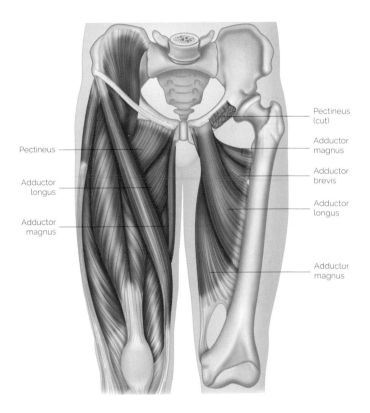

Pectineus

Adductor longus

Adductor magnus

Pectineus (cut)

Adductor magnus

Adductor brevis

Adductor longus

Adductor magnus

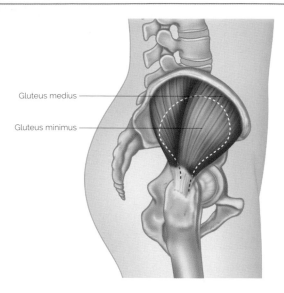

Gluteus medius

Gluteus minimus

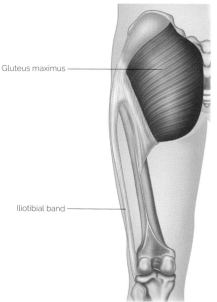

Gluteus maximus

Iliotibial band

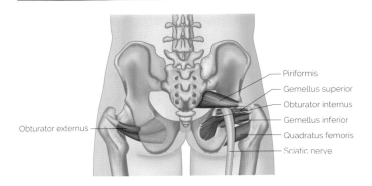

Obturator externus

Piriformis
Gemellus superior
Obturator internus
Gemellus inferior
Quadratus femoris
Sciatic nerve

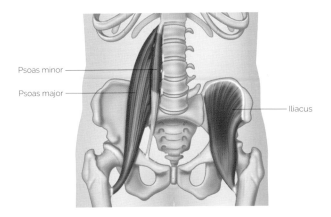

Psoas minor
Psoas major
Iliacus

⊃ Gluteus Maximus

Origin	Insertion	Nerve
Outer surface of ilium, behind posterior gluteal line and portion of bone superior and posterior to it. Adjacent posterior surface of sacrum and coccyx. Sacrotuberous ligament. Aponeurosis of erector spinae.	Deep fibers of distal portion: gluteal tuberosity of femur. Remaining fibers: iliotibial tract of fascia lata.	Inferior gluteal nerve L5, S1, 2.

⊃ Tensor Fasciae Latae

Origin	Insertion	Nerve
Anterior part of outer lip of iliac crest, and outer surface of anterior superior iliac spine.	Joins iliotibial tract just below level of greater trochanter.	Superior gluteal nerve L4, 5, S1.

⊃ Gluteus Medius

Origin	Insertion	Nerve
Outer surface of ilium between anterior and posterior gluteal lines.	Oblique ridge on lateral surface of greater trochanter of femur.	Superior gluteal nerve L4, 5, S1.

⊃ Gluteus Minimus

Origin	Insertion	Nerve
Outer surface of ilium between anterior and inferior gluteal lines.	Anterior border of greater trochanter.	Superior gluteal nerve L4, 5, S1.

⊃ Piriformis

Origin	Insertion	Nerve
Anterior surface of sacrum between anterior sacral foramina.	Superior border of greater trochanter of femur.	Branches from sacral nerves S1, 2.

⊃ Deep Lateral Hip Rotators

Origin	Insertion	Nerve
Obturator internus: inner surface of ischium, pubis, and ilium. Gemellus superior: ischial spine (lower posterior area of pelvis). Gemellus inferior: just below origin of gemellus superior. Quadratus femoris: lateral edge of ischial tuberosity (sitting bone).	Greater trochanter (top) of femur (except quadratus femoris which inserts just behind and below the others).	Obturator internus and gemellus superior: nerve to obturator internus, L5, S1, 2. Gemellus inferior and quadratus femoris: nerve to quadratus femoris, L4, 5, S1, (2).

⮑ **Pectineus**

Origin	Insertion	Nerve
Pectineal line and adjacent bone of pelvis.	Oblique line, from lesser trochanter to linea aspera of femur.	Femoral nerve L2, 3.

⮑ **Obturator Externus**

Origin	Insertion	Nerve
External surface of obturator membrane and adjacent bone.	Trochanteric fossa of femur.	Posterior division of obturator nerve L3, 4.

⮑ **Adductors**

Origin	Insertion	Nerve
Anterior part of the pubic bone (ramus). Adductor magnus also takes its origin from the ischial tuberosity.	Entire length of femur, along linea aspera and medial supracondylar line to adductor tubercle on medial epicondyle of femur.	Magnus: obturator nerve L2, 3, 4. Brevis: obturator nerve L2, 3. Longus: obturator nerve L2, 3, 4.

⮑ **Psoas Major**

Origin	Insertion	Nerve
Lateral surface of bodies of T12 and L1 to L5 vertebrae, transverse processes of the lumbar vertebrae, and the intervertebral discs between T12 and L1 to L5 vertebrae.	Lesser trochanter of femur.	Ventral rami of lumbar nerves L1–3 (psoas minor innervated from L1, 2).

⮑ **Iliacus**

Origin	Insertion	Nerve
Posterior abdominal wall (iliac fossa).	Lesser trochanter of femur.	Femoral nerve L2, 3.

Movements of the Hip Joint

Flexion	Extension	Abduction	Adduction	Lateral Rotation	Medial Rotation
Iliopsoas, rectus femoris, tensor fasciae latae, sartorius, adductor brevis, adductor longus, pectineus	Gluteus maximus, semitendinosus, semimembranosus, biceps femoris (long head), adductor magnus (ischial fibers)	Gluteus medius, gluteus minimus, tensor fasciae latae, obturator internus (in flexion), piriformis (in flexion)	Adductor magnus, adductor brevis, adductor longus, pectineus, gracilis, gluteus maximus (lower fibers), quadratus femoris	Gluteus maximus, obturator internus, gemelli, obturator externus, quadratus femoris, piriformis, sartorius, adductor magnus, adductor brevis, adductor longus	Iliopsoas (in initial stage of flexion), tensor fasciae latae, gluteus medius (anterior fibers), gluteus minimus (anterior fibers)

Muscles of the Thigh and Knee Joint

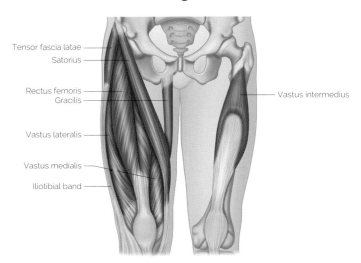

Tensor fascia latae

Satorius

Rectus femoris

Gracilis

Vastus lateralis

Vastus medialis

Iliotibial band

Vastus intermedius

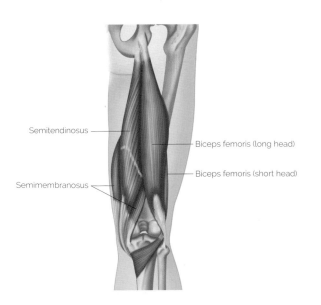

Semitendinosus

Semimembranosus

Biceps femoris (long head)

Biceps femoris (short head)

➲ Hamstrings

Origin	Insertion	Nerve
Ischial tuberosity (sitting bone). Biceps femoris also originates from the back of the femur.	Semimembranosus: groove and adjacent bone on medial and posterior surface of medial tibial condyle. Semitendinosus: upper medial surface of shaft of tibia. Biceps femoris: head of fibula.	Sciatic nerve L5, S1, 2.

➲ Gracilis

Origin	Insertion	Nerve
Line on the external surfaces of the body of the pubis. Inferior ramus of pubis. Ramus of the ischium.	Upper part of medial surface of shaft of tibia.	Obturator nerve L2, 3.

➲ Sartorius

Origin	Insertion	Nerve
Anterior superior iliac spine and the area immediately below it.	Upper part of medial surface of tibia, near anterior border.	Femoral nerve L2, 3.

➲ Quadriceps

Origin	Insertion	Nerve
Rectus femoris: anterior head: anterior inferior iliac spine; posterior head: groove above acetabulum (on ilium). Vasti group: upper half of shaft of femur.	Quadriceps femoris tendon.	Femoral nerve L2, 3, 4.

➲ Popliteus

Origin	Insertion	Nerve
Lateral condyle of femur.	Upper part of posterior surface of tibia, superior to soleal line.	Tibial nerve L4, 5, S1.

Movements of the Knee Joint

Flexion	Extension	Medial Rotation of Tibia on Femur	Lateral Rotation of Tibia on Femur
Hamstrings, gastrocnemius, plantaris, sartorius, gracilis, popliteus	Quadriceps femoris	Popliteus, semitendinosus, semimembranosus, sartorius, gracilis	Biceps femoris

Muscles of the Leg and Ankle Joint

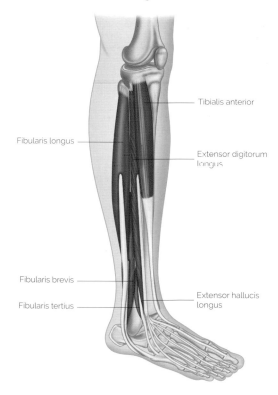

Tibialis anterior

Fibularis longus

Extensor digitorum longus

Fibularis brevis

Fibularis tertius

Extensor hallucis longus

➲ Tibialis Anterior

Origin	Insertion	Nerve
Lateral condyle of tibia. Upper half of lateral surface of tibia. Interosseous membrane.	Medial and plantar surface of medial cuneiform bone. Base of first metatarsal.	Deep fibular nerve L4, 5.

➲ Extensor Digitorum Longus

Origin	Insertion	Nerve
Lateral condyle of tibia. Upper two-thirds of medial surface of fibula.	Along dorsal surface of the four lateral toes. Each tendon divides, to attach to bases of middle and distal phalanges.	Deep fibular nerve L5, S1.

➲ Extensor Hallucis Longus

Origin	Insertion	Nerve
Middle half of medial surface of fibula and adjacent interosseous membrane.	Base of distal phalanx of great toe.	Deep fibular nerve L5, S1.

➲ Fibularis Tertius

Origin	Insertion	Nerve
Lower third of medial surface of fibula.	Dorsal surface of base of fifth metatarsal.	Deep fibular nerve L5, S1.

➲ Fibularis Longus and Fibularis Brevis

Origin	Insertion	Nerve
Fibularis longus: upper two-thirds of lateral surface of fibula. Lateral condyle of tibia. Fibularis brevis: lower two-thirds of lateral surface of fibula.	Fibularis longus: lateral side of medial cuneiform. Base of first metatarsal. Fibularis brevis: lateral side of base of fifth metatarsal.	Superficial fibular nerve L5, S1, 2.

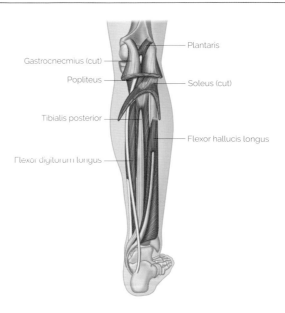

Plantaris

Gastrocnecmius (cut)

Popliteus

Soleus (cut)

Tibialis posterior

Flexor hallucis longus

Flexor digitorum longus

➲ Gastrocnemius

Origin	Insertion	Nerve
Medial head: popliteal surface of femur above medial condyle. Lateral head: lateral condyle and posterior surface of femur.	Posterior surface of calcaneus (via calcaneus—a fusion of the tendons of gastrocnemius and soleus).	Tibial nerve S1, 2.

➲ Soleus

Origin	Insertion	Nerve
Posterior surfaces of head of fibula and upper third of body of fibula. Soleal line and middle third of medial border of tibia. Tendinous arch between tibia and fibula.	Posterior surface of calcaneus (via calcaneus—a fusion of the tendons of gastrocnemius and soleus).	Tibial nerve L5, S1, 2.

⊃ **Plantaris**

Origin	Insertion	Nerve
Lower part of lateral supracondylar ridge of femur and adjacent part of its popliteal surface. Oblique popliteal ligament of knee joint.	Posterior surface of calcaneus (via calcaneus—a fusion of the tendons of gastrocnemius and soleus).	Tibial nerve S1, 2.

⊃ **Flexor Digitorum Longus**

Origin	Insertion	Nerve
Medial part of posterior surface of tibia, below soleal line.	Bases of distal phalanges of second through fifth toes.	Tibial nerve S2, 3.

⊃ **Flexor Hallucis Longus**

Origin	Insertion	Nerve
Lower two-thirds of posterior surface of fibula. Adjacent interosseous membrane.	Base of distal phalanx of great toe.	Tibial nerve S1, 2.

⊃ **Tibialis Posterior**

Origin	Insertion	Nerve
Lateral part of posterior surface of tibia. Upper two-thirds of posterior surface of fibula. Most of interosseous membrane.	Tuberosity of navicular. By fibrous expansions to sustentaculum tali, three cuneiforms, cuboid, and bases of second, third, and fourth metatarsals.	Tibial nerve L4, 5.

Movements of the Ankle Joint

Inversion	Eversion	Dorsiflexion	Plantar Flexion
Tibialis anterior, tibialis posterior	Fibularis longus, fibularis brevis, fibularis tertius	Tibialis anterior, extensor hallucis longus, extensor digitorum longus, fibularis tertius	Gastrocnemius, plantaris, soleus, tibialis posterior, flexor hallucis longus, flexor digitorum longus, fibularis longus, fibularis brevis

Muscles of the Foot

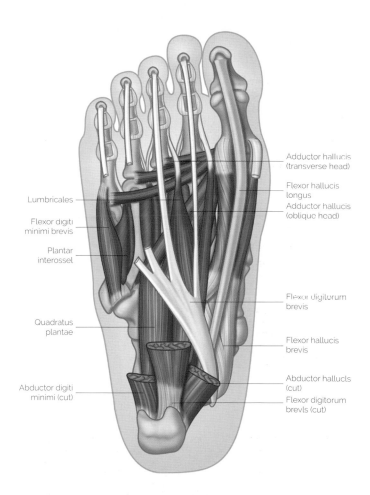

Adductor hallucis (transverse head)

Flexor hallucis longus

Adductor hallucis (oblique head)

Lumbricales

Flexor digiti minimi brevis

Plantar interossei

Flexor digitorum brevis

Quadratus plantae

Flexor hallucis brevis

Abductor hallucis (cut)

Abductor digiti minimi (cut)

Flexor digitorum brevis (cut)

⊃ **Abductor Hallucis**

Origin	Insertion	Nerve
Tuberosity of calcaneus. Flexor retinaculum. Plantar aponeurosis.	Medial side of base of proximal phalanx of great toe.	Medial plantar nerve L4, 5, S1.

⊃ **Flexor Digitorum Brevis**

Origin	Insertion	Nerve
Tuberosity of calcaneus. Plantar aponeurosis. Adjacent intermuscular septa.	Middle phalanges of second to fifth toes.	Medial plantar nerve L4, 5, S1.

⊃ **Abductor Digiti Minimi**

Origin	Insertion	Nerve
Tuberosity of calcaneus. Band of connective tissue connecting calcaneus with base of fifth metatarsal.	Lateral side of base of proximal phalanx of fifth toe.	Lateral plantar nerve S1, 2, 3.

⊃ **Quadratus Plantae**

Origin	Insertion	Nerve
Medial head: medial surface of calcaneus. Lateral head: lateral border of inferior surface of calcaneus.	Lateral border of tendon of flexor digitorum longus.	Lateral plantar nerve S1, 2, 3.

⊃ **Lumbricals**

Origin	Insertion	Nerve
Tendons of flexor digitorum longus.	Medial side of base of proximal phalanges of second to fifth toes and corresponding extensor hoods.	Lateral three lumbricals: lateral plantar nerve S2, 3. First lumbrical: medial plantar nerve S2, 3.

⊃ Flexor Hallucis Brevis

Origin	Insertion	Nerve
Medial part of plantar surface of cuboid bone. Adjacent part of lateral cuneiform bone. Tendon of tibialis posterior.	Medial part: medial side of base of proximal phalanx of great toe. Lateral part: lateral side of base of proximal phalanx of great toe.	Medial plantar nerve S1, 2.

⊃ Adductor Hallucis

Origin	Insertion	Nerve
Oblique head: bases of second to fourth metatarsals; sheath of fibularis longus tendon. Transverse head: plantar metatarsophalangeal ligaments of third to fifth toes; transverse metatarsal ligaments.	Lateral side of base of proximal phalanx of great toe.	Lateral plantar nerve S2, 3.

⊃ Flexor Digiti Minimi Brevis

Origin	Insertion	Nerve
Sheath of fibularis longus tendon. Base of fifth metatarsal.	Lateral side of base of proximal phalanx of little toe.	Lateral plantar nerve S2, 3.

⊃ Dorsal Interossei

Origin	Insertion	Nerve
Adjacent sides of metatarsal bones.	Extensor hoods and bases of proximal phalanges of second to fourth toes.	Lateral plantar nerve S2, 3.

⊃ Plantar Interossei

Origin	Insertion	Nerve
Bases and medial sides of third to fifth metatarsals.	Extensor hoods and bases of proximal phalanges of same toes.	Lateral plantar nerve S2, 3.

➲ **Extensor Digitorum Brevis and Extensor Hallucis Brevis**

Origin	Insertion	Nerve
Superolateral surface of calcaneus.	Extensor digitorum brevis: lateral sides of tendons of extensor digitorum longus to second to fourth toes. Extensor hallucis brevis: base of proximal phalanx of great toe.	Deep fibular nerve S1, 2.

Movements of the Intertarsal Joints

Inversion	Eversion	Other Movements
Tibialis anterior, tibialis posterior	Fibularis tertius, fibularis longus, fibularis brevis	Sliding movements, which allow some dorsiflexion, plantar flexion, abduction, and adduction, are produced by the muscles acting on the toes. Tibialis anterior, tibialis posterior, and fibularis tertius are also involved.

Movements of the Metatarsophalangeal Joints of the Toes

Flexion	Extension	Abduction and Adduction
Flexor hallucis brevis, flexor hallucis longus, flexor digitorum longus, flexor digitorum brevis, flexor digiti minimi brevis, lumbricales, interossei	Extensor hallucis longus, extensor hallucis brevis (great toe), extensor digitorum brevis, extensor digitorum longus	Abductor hallucis, adductor hallucis, interossei, abductor digiti minimi

Movements of the Interphalangeal Joints of the Toes

Flexion	Extension
Flexor hallucis longus, flexor digitorum brevis (proximal joint only), flexor digitorum longus	Extensor hallucis longus, extensor digitorum brevis (not in great toe), extensor digitorum longus, lumbricales

Resources

Alter, M.J. 1998. *Sport Stretch: 311 Stretches for 41 Sports*, Champaign, IL: Human Kinetics.

Anderson, D.M. (chief lexicographer) 2003. *Dorland's Illustrated Medical Dictionary*, 30th edn, Philadelphia, PA: Saunders.

Bartelink, D.L. 1957. The role of abdominal pressure in relieving the pressure on the lumbar intervertebral discs. *Journal of Bone and Joint Surgery* 39-B, 718.

Biel, A. 2001. *Trail Guide to the Body*, 2nd edn, Boulder, CO: Books of Discovery.

Bumke, O. & Foerster, O. (eds) 1936. *Handbuch der Neurologie*, Vol. V, Berlin: Julius Springer.

Clemente, C.M. (ed.) 1985. *Gray's Anatomy of the Human Body*, 30th edn, Philadelphia, PA: Lea & Febiger.

DeJong, R.N. 1967. *The Neurological Examination*, 3rd edn, New York: Harper & Row.

Fuller, G.N. & Burger, P.C. 1990. Nervus terminals (cranial nerve zero) in the adult human. *Clin Neuropathol* 9(6), 279–283.

Gracovetsky, S. 1988. *The Spinal Engine*. New York: Springer-Verlag Wein.

Haymaker, W. & Woodhall, B. 1953. *Peripheral Nerve Injuries*, 2nd edn, Philadelphia, PA: W.B. Saunders Co.

Hodges, P.W. & Richardson, C.A. 1997. Feedforward contraction of transversus abdominis is not influenced by direction of arm movement. *Experimental Brain Research* 114(2), 362–370.

Huijing, P.A. & Baan, G.C. 2001. Extramuscular myofascial force transmission within the rat anterior tibial compartment: Proximodistal differences in muscle force. *Acta Physiologica Scandinavica* 173(3), 297–311.

Huxley, H. & Hanson, J. 1954. Changes in the cross-striations of muscle during contraction and stretch and their structural interpretation. *Nature* 173(4412), 973–976.

Kendall, F.P. & McCreary, E.K. 1983. *Muscles, Testing & Function*, 3rd edn, Baltimore, MD: Williams & Wilkins.

Lawrence, M. 2004. *Complete Guide to Core Stability*, London: A&C Black.

Levin, S.M. 2002. The tensegrity-truss as a model for spine mechanics. *Journal of Mechanics in Medicine and Biology* 2(3&4), 375–388.

Masi, A.T. & Hannon, J.C. 2008. Human resting muscle tone (HRMT): Narrative introduction and modern concepts. *Journal of Bodywork and Movement Therapies* 12(4), 320–332.

Myers, T.W. 2014. *Anatomy Trains: Myofascial Meridians for Manual and Movement Therapists*, 3e, Edinburgh: Elsevier.

Norris, C.M. 1997. *Abdominal Training*, London A&C Black.

Romanes, G.J. (ed.) 1972. *Cunningham's Textbook of Anatomy*, 11th edn, London: Oxford University Press.

Schade, J.P. 1966. *The Peripheral Nervous System*, New York: Elsevier.

Spalteholz, W. (date unknown). *Hand Atlas of Human Anatomy*, Vols II and III, 6th edn, London: J.B. Lippincott.

Tortora, G. 1989. *Principles of Human Anatomy*, 5th edn, New York: Harper & Row.

Index